Climate Change Policy in Japan

Amidst growing environmental concerns worldwide, Japan is seen as particularly vulnerable to the effects of a changing climate. This book considers Japan's response to the climate change problem from the late 1980s up to the present day, assessing how the Japanese government's policy-making process has developed over time. From the early days of climate change policy in Japan, through the United Nations Framework Convention on Climate Change conferences and Kyoto Protocol, right up to the 2015 negotiations, the book examines the environmental, economic, and political factors that have shaped policy. As the 2015 Conference of the Parties to the United Nations Framework Convention on Climate Change projects forward beyond 2020, the book concludes by analyzing how Japan has placed itself in the global climate change debate and how the country might and should respond to the problem in the future, based on the findings from accumulated history.

Yasuko Kameyama is Deputy Director at the Center for Social and Environmental Systems Research, National Institute for Environmental Studies (NIES), Japan.

Routledge Studies in Asia and the Environment

The role of Asia will be crucial in tackling the world's environmental problems. The primary aim of this series is to publish original, high quality, research-level work by scholars in both the East and the West on all aspects of Asia and the environment. The series aims to cover all aspects of environmental issues, including how these relate to economic development, sustainability, technology, society, and government policies; and to include all regions of Asia.

Climate Change Policy in Japan

From the 1980s to 2015

Yasuko Kameyama

Routledge
Taylor & Francis Group

LONDON AND NEW YORK

First published 2017 by Routledge

2 Park Square, Milton Park, Abingdon, Oxfordshire OX14 4RN
52 Vanderbilt Avenue, New York, NY 10017

Routledge is an imprint of the Taylor & Francis Group, an informa business

First issued in paperback 2019

British Library Cataloguing in Publication Data
A catalogue record for this book is available from the British Library

Library of Congress Cataloging in Publication Data
Names: Kameyama, Yasuko.
Title: Climate change policy in Japan, from the 1980s to 2015 / Yasuko Kameyama.
Description: Abingdon, Oxon ; New York, NY : Routledge, 2016. | Series: Routledge studies in Asia and the environment ; 5 | Includes bibliographical references and index.
Identifiers: LCCN 2016023537 | ISBN 9781138838598 (hardback) | ISBN 9781315733920 (ebook)
Subjects: LCSH: Climatic changes—Japan. | Climatic changes—Political aspects—Japan. | Climatic changes—Government policy—Japan.
Classification: LCC QC903 .K1945 2016 | DDC 363.738/745610952—dc23
LC record available at https://lccn.loc.gov/2016023537

ISBN: 978-1-138-83859-8 (hbk)
ISBN: 978-0-367-18672-2 (pbk)

Typeset in Times
by Apex CoVantage, LLC

Contents

Tables and figures

Tables

Figures

Abbreviations

ACE	Actions for Cool Earth
ADP	Ad Hoc Working Group on Durban Platform
AGBM	Ad Hoc Group on the Berlin Mandate
AIJ	Activities Implemented Jointly
AIM	Asia-Pacific Integrated Model
ANRE	Agency for Natural Resources and Energy
AOSIS	Alliance of Small Island States
APP	Asia-Pacific Partnership for Clean Development and Climate
AR4	The 4th Assessment Report (of the IPCC)
AWG-KP	Ad Hoc Working Group on the Kyoto Protocol
AWG-LCA	Ad Hoc Working Group on Long-Term Cooperative Action under the Convention
BAU	business-as-usual
BOCM	Bilateral Offset Credit Mechanism
CASA	Citizens' Alliance for Saving the Atmosphere and the Earth
CCX	Chicago Climate Exchange
CDM	Clean Development Mechanism
CER	Certified Emission Reductions
CGE	Computable General Equilibrium
CMP	Conference of the Parties serving as the Meeting of the Parties
COP	Conference of the Parties
CFC	Chlorofluorocarbon
CH_4	Methane
CO_2	Carbon Dioxide
DPJ	Democratic Party of Japan
EA	Environment Agency
ETS	Emissions Trading Scheme
EU	European Union
G8	Group of Eight
GCF	Green Climate Fund
GDP	Gross Domestic Product
GEA	Global Environment Action
GHG	Greenhouse Gas

GNT	Geneva Negotiating Text
HFC	Hydrofluorocarbon
ICLEI	International Council for Local Environmental Initiatives
IEA	International Energy Agency
IEEJ	Institute for Energy Economics, Japan
INC	Intergovernmental Negotiation Committee
INDC	Intended Nationally Determined Contribution
IPCC	Intergovernmental Panel on Climate Change
JCM	Joint Crediting Mechanism
JI	Joint Implementation
JMA	Japan Meteorological Agency
Japan-CLP	Japan Climate Leaders' Partnership
JPY	Japanese Yen
JUSCANZ	A negotiating group consisting of Australia, Canada, Japan, New Zealand, and the United States
LDCs	Least-Developed Countries
LDP	Liberal Democratic Party
LNG	Liquified Natural Gas
LULUCF	Land Use, Land-Use Changes, and Forestry
MAC	marginal abatement cost
MAFF	Ministry of Agriculture, Forestry and Fisheries
MEM	Major Economies Meeting on Energy Security and Climate
METI	Ministry of Economy, Trade and Industry
MEXT	Ministry of Education, Culture, Sports, Science and Technology
MITI	Ministry of International Trade and Industry
MLIT	Ministry of Land, Infrastructure, Transport and Tourism
MOE	Ministry of the Environment
MOF	Ministry of Finance
MOFA	Ministry of Foreign Affairs
MRV	measurement, reporting and verification
N_2O	Nitrous Oxide
NIES	National Institute for Environmental Studies
NGO	nongovernmental organization
ODA	Overseas Development Aid
OECD	Organization for Economic Cooperation and Development
PFC	Perfluorocarbon
PKO	Peace Keeping Operations
PMO	Prime Minister's Office
PV	Photovoltaic
QELROs	Quantified Limitation and Reduction Objectives
REDD+	Reducing Emissions from Deforestation and Forest Degradation in Developing Countries
RITE	Research Institute of Innovative Technology for the Earth
SAR	The Second Assessment Report (of the IPCC)
SB	Subsidiary Body

SDPJ	Social Democratic Party of Japan
SF$_6$	Sulfur Hexafluoride
TEM	Technical Expert Meetings
TMG	Tokyo Metropolitan Government
UNCED	United Nations Conference of the Environment and Development
UNEP	United Nations Environment Programme
UNFCCC	United Nations Framework Convention on Climate Change
UNSC	United Nations Security Council
WCED	World Commission on Environment and Development
WMO	World Meteorological Organization
WSSD	World Summit on Sustainable Development
WWF	World Wide Fund for Nature

Acknowledgments

This book is the product of the many days I spent over the last 28 years studying at, and working for, the National Institute for Environmental Studies (NIES), conducting studies on climate change policy, participating in a series of multilateral negotiations under the United Nations Framework Convention on Climate Change as a member of the Japanese delegation, and being involved in decision-making for the national government, as well as local governments, regarding climate change mitigation and adaptation policies. Over this long time span, the scientific evidence on climate change has gradually solidified. In a manner parallel to the situation with climate change, Earth is surely on a trajectory toward global warming – a phenomenon that has never before been experienced by humans.

The long time period of evolution of this book allowed me to meet countless people who were concerned in one way or another about the climate change problem. I am indebted to all of these individuals for the precious experience, conversations, and knowledge they shared with me. Although I cannot list all of their names here, I would like to mention a few.

First and foremost, I am deeply grateful to Tsuneyuki Morita, who guided me to enter life as an academic expert on environmental policy and politics. He passed away in 2003 at only 53, but in his colleagues' minds his passion for climate change science and policy interface remains sound and solid. Other members of NIES – Shuzo Nishioka, Mikiko Kainuma, Toshihiko Masui, Kiyoshi Takahashi, Yasuaki Hijioka, and Izumi Kubota, who were the founders of climate change policy modeling at NIES – also inspired me with their quantitative modeling simulations.

My experiences in Japanese government delegations, as well as at various council meetings, were provided through the courtesy of officers of the Japanese Ministry of the Environment, including Hironori Hamanaka, Kazuhiko Takemoto, Hikaru Kobayashi, Takahiro Hiraishi, Katsunori Suzuki, Satoshi Tanaka, and Shigemoto Kajihara.

There are many Japanese experts with whom I enjoyed working. I particularly thank Yukari Takamura, Kentaro Tamura, Hiroshi Ohta, Isao Sakaguchi, and Norichika Kanie for sharing their wonderful expert knowledge and wisdom.

Comments were helpful in improving the manuscript at the final stage. My sincere thanks go to Detlef Sprinz, Miranda Schreurs, Rie Watanabe, and Atsushi Ishii for their invaluable comments and advice. Any remaining faults in the text are all mine.

I have received financial support from the Global Environmental Research Fund (2–1501) of the Ministry of the Environment, as well as a Science Research Grant from the Ministry of Education, Culture, Sports, Science and Technology.

Finally, my deep thanks go to my two children, Yuji and Haruka, who have put up with their mother working to all hours on her PC at home in the evenings! Their generation is the one that will have to live in a world with a changed climate in the coming years.

1 Framing Japan's response to climate change

Overview

Haru no umi hinemosu notari notari kana
(Spring ocean / swaying gently / all day long)

Buson Yosa

The Yosa poem is a renowned haiku from the eighteenth century. The beauty of four distinct seasons has long been at the heart of Japanese culture. In haiku, poets attempt to express their deepest emotions by describing nature in simple and beautiful language in the style of a short poem with three lines consisting of five, seven, and five syllables. In Buson Yosa's haiku, he was expressing a typical day in spring with calm waves and warm sunshine.

Spring 2015 in Japan was not the spring Buson described three centuries ago. Early April was as cold as midwinter, with snowfall in Tokyo. Then strong winds and rainfalls hit record highs in scattered areas around Japan. The cold spring was followed by clear days in early May, with temperatures reaching record highs; the amount of rainfall was only about 60% of the long-term average for May.

Summer was also very irregular, starting with extremely hot temperatures for three weeks from late July to mid-August. Temperatures then dropped to below 20 °C, which is much cooler than usual and would not usually be observed until mid-October in Japan. The cool spell was followed by an unprecedented amount of rainfall in the northern Kanto area in early September, which led to serious flooding across a wide area. Most Japanese feel that the climate is changing, and many of them are aware of the terms "climate change" and "global warming." Despite these extreme weather patterns and a basic knowledge about of climate change, there is little enthusiasm, if any, among the Japanese people and government to start taking actions to address the climate change problem.

This chapter gives an overall introduction to the book. It explains the aim of the book and where the book stands among the large amount of literature related to climate change policy-making in general, focusing on the spectrum across which past literature has primarily concentrated. It also reviews how other published works have dealt with climate change policy-making in Japan.

Aim of this book

Few will deny that climate change has now become one of the most serious global environmental problems and that it requires a global solution (Gore 1993). Climate change alters not only average global temperature but also a series of climate-related events, such as precipitation and wind patterns. These changes have other environmental and ecological consequences, including desertification and loss of biological diversity, all of which directly or indirectly affect living conditions today and in the future.

A global response is required to mitigate climate change, and it has occurred mainly at the multilateral level. The United Nations Framework Convention on Climate Change (UNFCCC) was adopted in 1992 and entered into force in 1994. The Kyoto Protocol was adopted in 1997 and entered into force in 2005. Ongoing negotiations for the time periods beyond the first commitment period of the Kyoto Protocol (2008–2012) resulted in a political declaration known as the Copenhagen Accords (agreed upon in 2009), which paved a way for countries to move forward, mostly voluntarily, through 2020. The contents of the Copenhagen Accord were anchored to the UNFCCC process as Conference of the Parties (COP) decisions in 2010 in the Cancun Agreement. A new round of negotiations was initiated by the Durban Platform in 2011, which engaged all countries to take part in the newly agreed-upon outcome; the aim was to conclude the agreement in 2015 at COP21. Japan has consistently been engaged in these multilateral negotiations from the early stages in the late 1980s to the present.

Japan's emission of greenhouse gases (GHGs) is not negligible. With about 2% of the world's population, Japan has been responsible for about 3% to 4% of global emissions. It ranked fourth in GHG emissions in the 1980s, although that ranking has gradually decreased in the past two decades because of growing emissions in some emerging economies. Even though its overall rank has decreased, Japan has been one of the world's major economies since the 1980s, and other countries expect it to play a major role in international affairs, especially regarding climate change.

It has generally been difficult to fully understand how and why Japan has made certain decisions concerning climate change, particularly for many non-Japanese audiences. This is true for several reasons. First and foremost, analyses of Japan's foreign policy have generally tended to emphasize Japan's uniqueness. For many observers, "Japan appears anomalous, if not aberrant or abnormal, in terms of its international behaviour" (Hook et al. 2012: 68). Japan's foreign policy has also been traditionally perceived as reactive, whereas that of other industrialized countries is considered more or less proactive (Calder 1988; Inoguchi 1991). Similar observations can be made for Japan's response to climate change. It is not clear to outside observers how key decision-making individuals and other stakeholders perceive climate change as an issue, how they use logic to arrive at decisions, and why some key factors in other countries do not exert the same level of influence in Japan. There is also little transparency with respect to the process by which the major players in the decision-making process consolidated their final decisions, including the information chosen to be considered and the reasons why some elements of the climate change problem have been ignored.

Also, most relevant literature and publicly available information on the climate change debate in Japan is written in Japanese, which restricts access to much of the non-Japanese audience. Finally, most Japanese people seem to perceive climate change as an economic and energy issue, rather than as an environmental, ethical, development, or diplomatic issue. Thus, there is a kind of common understanding among Japanese people when discussing climate change, even though the Japanese dialogue does not always seem to properly fit into the climate change policy puzzle in the multilateral arena. This point of view will be explored later in the book in following chapters.

The primary purpose of this book is, therefore, to examine the trajectory of Japan's decision-making processes regarding responses to the climate change problem. It focuses particularly on high-ranking politicians to examine how they approached climate change and how they perceived climate change in relation to other political, economic, and social issues. Because climate change is related to many other national issues, politicians may have dealt with issues that were not strictly related to climate change per se, but that were nonetheless relevant.

A country's decisions regarding climate change are affected by many factors, such as scientific findings and economic conditions. Key influential factors can change over time, depending on specific conditions. In the rest of this chapter, I examine key factors that have influenced Japan's decision-making on climate change policies in the past three decades and note how these factors have evolved over time. I also identify the factors that are unique to Japan and those that are common across countries.

What is the climate change problem?

The climate change mechanism is a phenomenon that can be explained by physical processes. The climate change problem, however, is not merely a physical phenomenon; it is a problem that entails political, economic, and social dimensions. The problem, therefore, can be framed in a variety of ways depending on which dimension is being regarded as the central concern. To investigate Japan's response to climate change, it is necessary to delineate suitable dimensions in the Japanese context of climate change. Generally, climate change dimensions can be categorized in three types: environmental, economic, and foreign policy.

Climate change as an environmental issue

Climate change is no doubt a global environmental problem. It is a mechanism through which people's use of fossil fuel energy, destruction of forests, and emission of other GHGs such as fluorocarbon gases has led to an increase in concentration of GHGs in the atmosphere. Through this increase, more heat is trapped in the atmosphere, resulting in gradual warming in terms of global average temperature and more severe weather patterns at the regional and local levels.

Scientists have helped formalize the founding basis for multilateral cooperation to tackle climate change. The Intergovernmental Panel on Climate Change (IPCC) has published five assessment reports thus far, and their findings have shown continuous upward trends of atmospheric concentration of GHGs, a global temperature rise, and a higher frequency of extreme weather events (IPCC 2015). From an environmental conservation perspective, the aim is to restrict the temperature increase to a long-term goal of 2 °C, or even of 1.5 °C, a target that was agreed upon in 2010 as an element of the Cancun Agreement (UNFCCC 2010) and in 2015 by the Paris Agreement (UNFCCC 2015). To reach this goal, it is estimated that total global GHG emissions need to be reduced by half of those in the 1990s (G8 Hokkaido Toyako Summit 2008). The mean global temperature has already risen by about 0.85 °C, suggesting the difficulty involved in reaching this goal within a set timeframe (IPCC 2015).

Meanwhile, there are some climate change "deniers" who state that, although climate change may be occurring, it is not because of anthropogenic GHG emissions or that it is not a serious problem. Debates between supporters and deniers of climate change science often include discussion of the political and economic dimensions of the problem (Bradley 2011; Giddens 2009; Jacques 2012).

Japan's climate is already changing. The annual average temperature in Japan in 2014 was 0.14 °C higher than the annual average from 1989 to 2014. In addition, the long-term trend shows a temperature increase of 1.14 °C in the last century (Figure 1.1) (Meteorological Agency 2015). Although the amount of rainfall tends to vary from year to year, the occurrence of extreme events (i.e., too much or too little rain) has been increasing in the last decade. Although it does not directly refer to climate change, the latest White Paper from the Fire and Disaster Management Agency (FDMA) notes an increasing frequency of tremendously concentrated rainfall in limited areas and powerful typhoons and tornados, leading to increased damage incurred by these extreme weather patterns (FDMA 2014). For example, more than 40,000 people have been hospitalized for heat stroke each year since 2010. The number had been less than 10,000 in the previous years.

Climate change as an economic/energy issue

Among a variety of types of GHGs identified by scientists, carbon dioxide (CO_2) is the main target that needs to be addressed to avoid dangerous consequences of global warming on our climate and ecosystem. Reduction of CO_2 emissions is also the core issue of climate change problem in terms of the economic dimension. There are assertions that a reduction in energy use would hamper economic growth (White House 2001). The term "burden sharing," an expression often used in discussions on ways to set emission reduction targets across countries (particularly in the early years of negotiations), clearly indicates people's recognition that emission reduction is a burden (Ringius et al. 2002). This debate very much depends upon how "economy" and "cost" are defined. In terms of growth of gross domestic product (GDP), there is generally a positive relationship between

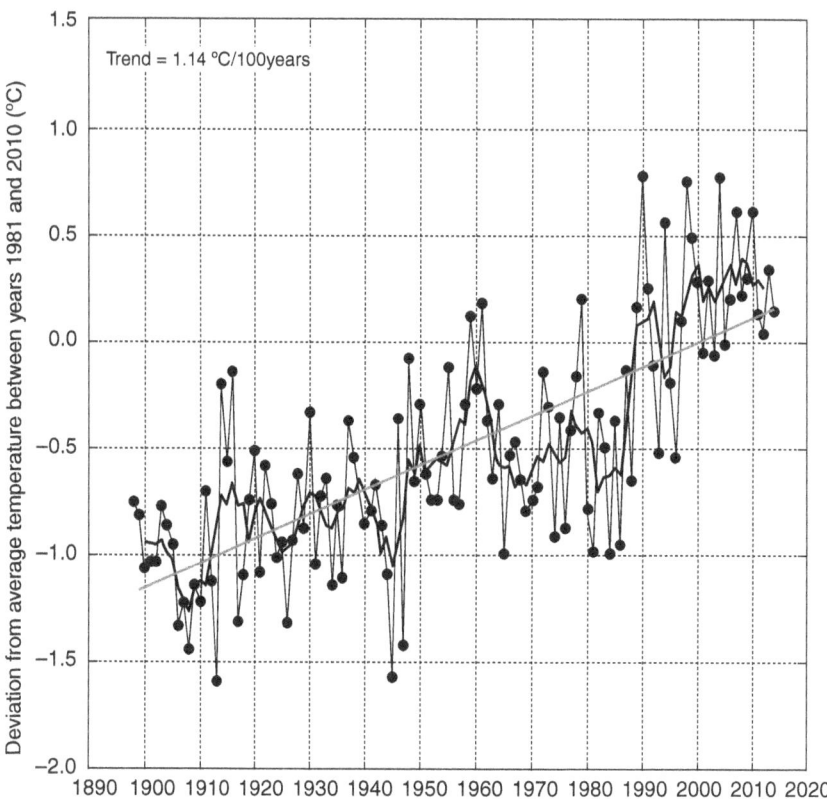

Figure 1.1 Annual average temperature and long-term trend

growth in CO_2 emissions and GDP, but "decoupling" between the two has also been observed in some countries and time periods (Grubb 2014). Decoupling tends to occur in countries where the energy price is relatively high and where the diffusion of new technology has occurred. The economic cost of climate change mitigation itself is a complicated notion. At least some extra expenditure would have to be made if a new, less-carbon-intensive technology were to be installed all at once in a short timeframe. This is perceived as an economic cost or burden in the short term. For example, for many individuals, an increase in the price of gasoline or electricity would not be welcome, because they will perceive this as an increased household "cost."

Many studies have calculated the total economic cost of reducing GHG emissions worldwide (IPCC 2014: Chapter 6), but the estimates are wide ranging, primarily because of the diversity of various underlying assumptions. Examples include the prospects for future population and economic growth with and without climate change mitigation policies, the price of crude oil, possible utilization of innovative technologies, and the availability of renewable and nonrenewable energy resources.

Decreased energy consumption, however, also saves money, and many energy-saving actions have yet to be implemented. From this perspective, expenditures required to save energy should be considered as investments that may end up saving money in the long term. The benefit of cost saving through the reduced use of energy increases when the prices of fossil fuel resources increase. From a government's fiscal policy perspective, reducing subsidies for fossil fuels is a climate change mitigation policy that actually reduces government expenditure (IEA 2015: 81). Many developing countries used to operate energy subsidies to support the poor, but the use of subsidies has been re-evaluated in many countries in recent years.

From an industry perspective, many fossil fuel industries such as the coal and oil industries and many other energy-intensive industries such as the iron and steel, aluminum, and cement sectors are likely to be negatively affected economically by GHG emission reduction policies. These industries are, in many cases, politically powerful and have influenced national governments' behavior on climate change policy in many countries (Gelbspan 2005). On the other hand, there are emerging industries such as those related to renewable energy and low-carbon technologies that could benefit by ambitious emission-reduction policies. The notions of "green growth" or a "green economy" aim to express the concept that sustainable economic activity and environmental conservation can be achieved simultaneously (Stern 2013). In general, the overall impact of climate change mitigation policies on a country's economy needs to be viewed from many perspectives.

Climate change as a diplomatic/foreign-policy issue

Climate change is a long-term policy problem that lasts at least one generation, exhibits scientific uncertainty, and engenders a public goods aspect at the stage of problem generation and at the response stage, both at the global level (Sprinz 2009). When it comes to multilateral negotiations, it can be a contentious issue between rich and poor countries. Past CO_2 emissions are mostly from rich developed countries that burned fossil fuels. Today, many developing countries with emerging economies hope to become wealthy through increased industrialization, and they criticize the developed countries for the double standard of using their own past emissions to grow while requesting that the developing countries restrict their GHG emissions. The use of official development assistance in the fields of climate change mitigation and adaptation in the least developed countries is considered a precondition for these countries to be able to take action against climate change. From the developing countries' viewpoint, climate change is an issue of the right to development, justice, and equity (Pinguelli-Rosa and Munasinghe 2002; Sagar 2000; Shukla 1999; Tóth 1999).

From the perspective of diplomacy among wealthy countries, climate change and other environmental issues can be viewed as useful tools to exert leadership

(Underdal 1997; Young 1994). Environmental diplomacy could be seen as a type of "beauty contest," in which some countries are vying to be seen as clean or green as possible. For countries that do not have hard military power, making contributions to global environmental issues can lead to increased "soft power" (Nye 2004).

The timing of the emergence of global environmental issues in the late 1980s is not by accident. What had been called "high politics," such as those related to security matters, largely disappeared in the late 1980s because of the end of the Cold War. Then issues related to "low politics," including environmental and humanitarian agendas, became elevated to high politics. Politicians and government officials sensed the importance of climate change as a new international agenda that could also affect other fields and agendas (Sands 1992).

Who determines a country's position on climate change?

A long list of published works deals with decision-making processes and decision-making factors regarding countries' responses to environmental problems, particularly in the fields of comparative politics and public policy (Boehmer-Christiansen and Weidner 1995; Collier and Löfstedt 1997; Hajer 1995; Harrison and Sundstrom 2010; Jänicke and Weidner 1995; O'Riordan and Jäger 1996; Social Learning Group 2001; Steinberg and VanDeveer 2012; Vogel 1986). A large portion of the literature uses Western industrialized countries under democratic governance as case studies or as the given underlying community.

Official positions and decisions of democratic nation-states are formalized by national governments. In a way, the decision-making of countries can be considered a process of "corporatist consensus-seeking through elitist bargaining" (Opschoor and van der Straaten 1993). Members of government committees (Howlett and Ramesh 2003) and policy elites (Heinz et al. 1990) tend to be examined in studies of nation-states' consolidated decision-making in the realm of political science literature in general.

Relatively few studies in this field focus particularly on inter-ministerial or inter-departmental debates inside a bureaucracy. Rather, the scope of analysis usually perceives countries' decisions as a result of formal and informal consultations among various nonstate actors and stakeholders relevant to the issue (Hochstetler 2012; Kingdon 1995; Meadowcroft 2004). Nonstate actors in this case include scientific experts, environmental nongovernmental organizations (NGOs), interest groups (particularly industry and business groups), members of governmental committees, policy elites, and politicians. These actors form a kind of issue network (Heclo 1978), policy network (Marsh and Rhodes 1992; Rhodes 1990; Richardson and Jordan 1979), or coalition (Sabatier and Jenkins-Smith 1993; Smith 1993) to push their respective points of view. A country's decision can thus be interpreted as an outcome of interactions among interest groups in the country, the underlying rules of the interaction, and common values shared among the participating groups.

These subnational actors have different aims and norms, so that environmental issues must be framed in dimensions that differ from those discussed in the section above.

Political leaders

Political leaders in the modern party system are political entrepreneurs, "a team of men seeking to control the governing apparatus by gaining office in a duly constituted election" (Downs 1957). Their primary objective is reelection. Most political leaders expect to stay in power, so it can be assumed that they more or less reflect the preferences of some individuals and industry groups in order to gain their support at the time of elections (Urpelainen 2012).

Studies of politicians in the field of environmental studies have focused mainly on the emergence of Green Parties in the 1980s and 1990s (Alber 1989; Chandler and Siaroff 1986; Kitschelt 1993; Rüdig 1985). The proliferation of Green Parties was, however, observed predominantly in Western European countries, and, even in this limited region, traditional political parties regained power by incorporating environmental concerns into their own mainstream agendas (O'Neill 2012).

Rather than continuing to view green political parties as the key determinant group, more attention needs to be paid to individual political leaders and members of parliaments. They may want to be viewed as "green" to win votes from citizens with high levels of environmental awareness, or they may not want to be viewed as "green" to win votes from individuals related to energy-intensive industries. In countries like the United States with a stable two-party system, it is difficult to envision the emergence of a dominant Green Party. Nevertheless, there are surely individual politicians, such as former vice president Al Gore, who are known as active environmentalists.

The scientific community

Scientific findings and expert judgment are the fundamental bases of debates over any environmental policy. Meanwhile, a fundamental debate on the relationship between science and politics existed even before the establishment of the IPCC. Scientific research cannot always answer high-priority questions in terms of policy decisions with the precision, confidence, and timeliness that policymakers want (Weinberg 1972). Decision-makers therefore tend to use the best-available scientific findings to justify their respective views (Collingridge and Reeve 1986; Ezrahi 1980; Gieryn 1983; Weingart 1982; Zehr 2005).

Literature in the field of international relations has argued that scientific knowledge and expert judgment by an epistemic community can be influential in the domain of multilateral agreements (Haas 1989, 1992, 2007; Litfin 1994; Miller 2007; Miller and Edwards 2001; Parson 2003). This is also true of scientific communities in the national domain. The interplay between science and policy-making has been widely studied (Bryner 1993; Guston 1999; Jasanoff 1990, 2005; Latour 1987; Shackley and Wynne 1995). In the case of climate change, science or knowledge can affect decision-makers' perception in both directions, either for or against stringent GHG emission reductions. As can be found in the IPCC assessment reports, scientific findings on the mechanisms of climate change, future projections, plausible temperature increases, and the impact of climate change suggest that climate change should be mitigated "at a level that would prevent dangerous anthropogenic interference with the climate system" (UNFCCC Article 2). At the same time, there is a wide variety of studies on both the costs incurred by the adverse impacts of climate change and the costs required to reduce GHG emissions to avoid such damages, and some may suggest it may be more rational for nations to choose not to reduce their GHG emissions (Sprinz and Vaatoranta 1994).

Environmental NGOs

Similar to scientific experts, environmental NGOs also affect policy decisions on global environmental issues at both the international and the domestic levels. The transnational activities of environmental NGOs have been a focus of studies since the early 1990s (Betsill 2011; Betsill and Corell 2011; Conca and Lipschutz 1993; Doherty and Doyle 2006; Keck and Sikkink 1998; Princen and Finger 1994; Wapner 1996). Studies of these groups' domestic activities have also demonstrated how environmental NGOs have gained the power to influence national decisions on environmental issues (Desombre 2000; Dryzek et al. 2003; Inglehart 1982; Tarrow 1998). In the area of domestic policy-making, the NGOs' level of influence on decision-making varies from country to country. In some countries, particularly in the Asian region, NGOs have become more established when the form of governance has transitioned from central organization to a more diffused democratic style (Jancar-Webster 1998; Lee and So 1999).

Industry and business

Industry and the business community are among the most influential groups in countries' decision-making on environmental issues, both as transnational actors influencing multilateral level decisions (Schmidheiny 1992) and as influential domestic actors (Kraft and Kamieniecki 2007). In terms of the climate change problem, however, climate mitigation policies offer different costs and benefits to different industries. For example, many energy-intensive industries cannot avoid energy use in manufacturing their products (e.g., steel and aluminum). They would most likely have to reduce their level of production to reduce CO_2 emissions. These industries traditionally have been, and in many respects still are, the strongest advocates against domestic emission reduction policies. Meanwhile, some other energy-related industries have shifted their focus from fossil fuels to innovative low-carbon emission methods of operation to reduce emissions (Kolk and Levy 2001; Kolk and Pinkse 2005; Skjærseth and Skodvin 2003). Some industries also have taken the opportunity to gain economically by reducing CO_2 emissions or by publicizing their activities to consumers to create a positive image as a "green" company (Frankental 2001). Some emerging industries, such as those related to renewable energy, have welcomed countries' decisions to take positive action on emissions reductions.

The case of Japan

When it comes to Japan's decision-making and behavior on climate change and other environmental issues, researchers view the approaches taken quite differently from those used in Europe or North America.

Decision-making processes generally vary greatly from one country to another. When evaluating Japan's decision-making on climate change, one needs to account for the way in which Japan determines its positions. In general, experts in the field of Japanese politics view the country's decision-making as a tripartite power elite model (Mills 1956). The three elite groups – namely, the central bureaucracy, the ruling political party, and big business organizations – form an "iron triangle," sharing unofficial human networks and collaborating with each other to exclude other actors from political influence (Nester 1990). Most of literatures on Japan's decision making on climate change have applied this general model to a certain extent to explain the country's behavior.

Bureaucracy: power elites in ministries

The central bureaucracy has been seen as the dominant actor in Japan, with the ability to influence ultimate decisions (Johnson 1982; Pempel 1979). As Takashi Inoguchi (2012) wrote, the contemporary Japanese political system can be described as "bureaucratically led, mass-inclusionary pluralism." He continues,

> although the Japanese bureaucracy is strong, takes initiatives in many ways, and is proud that it takes into account as many of the preferences of the masses as possible, it is also a bureaucracy which is fragmented into ministries and agencies and functions in a political system which does not ensure that the various interests of these bodies will be melded into policy by the cabinet and prime minister.
>
> (Inoguchi 2012: 120)

The pronounced arguments concerning climate change policy-making in Japan is quite similar, with an emphasis on the persistent conflict among ministries, especially between the Ministry of the Environment (MOE; the Environment Agency [EA] until 2001) and the Ministry of Economy, Trade and Industry (METI; the Ministry of International Trade and Industry [MITI] until 2001) (Kameyama 2003; Kawashima 2000, 2001; Schreurs 2003; Tiberghien and Schreurs 2010). Compared with those of many other industrial democracies, Japan's decision-making owes much to bureaucratic processes. Politicians are advised and influenced by government officials, so their decisions are often tied to ministry interests.

Ministries' interests are often independent of each other. Basically, each ministry's interest is to increase its budget and staff. It is in MOE's interest to push the climate change agenda along with other major economic-social agendas in Japan. Climate change is a useful tool for developing MOE's administrative power over energy and fiscal policies, both basically under control of other ministries. It is in MITI's interest, however, to protect and promote activities related to Japanese industries and businesses. Securing energy resources is also in MITI's domain. It is in the Ministry of Foreign Affairs' (MOFA) interest to improve the image and standing of Japan when negotiating with foreign partners. The Ministry of Finance's interest is to increase the annual budget and decrease annual spending, as Japan has long been worsening its budgetary deficit. The Ministry of Agriculture, Forestry and Fisheries (MAFF) is responsible for forest-related policies, including carbon sequestration through afforestation and improved forest management as well as the mitigation of emissions from agriculture and fisheries. The Ministry of Land, Infrastructure, Transport and Tourism (MLIT) – which is responsible for energy efficiency standards for buildings, houses, and vehicles together with METI, as well as for GHG emissions reductions from aviation and marine transport – is anxious not to burden transportation industries by forcing reductions in energy consumption.

Other ministries – such as the Ministry of Education, Culture, Sports, Science and Technology (MEXT) and the Japan Meteorological Agency (JMA) – also have interests related to climate change from the perspective of monitoring, conducting research, and sharing scientific information on climate change mechanisms, the impact of climate change, and climate change mitigation and adaptation policies. With these different objectives in mind, ministries sometimes cooperate with each other, but at other times they are in conflict. Politicians become involved in decision-making only when the government faces a significant decision and when the conflicts among ministries seem to be unsolvable.

Asuka-Zhang (2003) investigated Japan's development assistance from the standpoint of climate change and concluded that use of public funds by the Japanese government was effective in mitigating GHG emissions at relatively low cost. The study also concluded that Japan's assistance programs will have a large impact on global GHG emission because of the large amount of emission growth that will be occurring in newly emerging economies in Asia, and in addition concluded that Japan needs to be aware of its role in terms of foreign policy, particularly in the Asian region.

Hattori (2007) evaluated Japan's climate change politics incorporating "norms" into the analysis. Focusing on various norms such as economic growth, energy efficiency, international contribution, and environmental protection, he argued that different norms were prioritized in different time periods by different ministries; this resulted in varying degrees of ambitiousness in the response to climate change. He also concluded that, by engaging in international action on climate change policy-making, non-state actors (including environmental NGOs and industries) learned how to take action independent of the government.

Political parties and leaders

The type of election system affects which politicians are more likely to be elected in national elections. According to Lijphart (1999), Japan used a "semi-proportional" system, under the medium-sized multimember district system, through 1995, and then switched to a system that combines a single-member-district plurality formula with proportional representation. The current system is thought to be beneficial for relatively powerful parties such as the Liberal Democratic Party (LDP), which can offer support to members in all regions of Japan.

The LDP has enjoyed the most stable one-party dominance of any advanced liberal democratic countries in the postwar era. Even so, the LDP has tended to follow "the norm of cross-party consensus building" (Pempel 1992). The head of the ruling party becomes Japan's prime minister, and thus the head of the LDP has occupied the position for more than five decades. The prime minister is formally vested with the power to choose ministers in the cabinet. The authority of the prime minister's position, however, is constrained by informal factors such as internal politics within the LDP, which aspires to maintain a peaceful balance between its various factions, bureaucracy and sectionalism, and opposition party resistance based on consensus-oriented rules and tactics in the Diet (Knoke et al. 1996: 233).

Most Japanese politicians, at least until recently, have relied heavily on ministerial officials for preparation of cabinet bills, budgets, and treaties submitted to the Diet by the cabinet. In addition, the Japanese ministries have developed a wide variety of "third sector" organizations – such as advisory councils, study groups, research consortia, public corporations, and foundations (Muramatsu and Tsujinaka 1992: 222–225) – that work to support the goals of the ministries. Not many Japanese political leaders in the past few decades have participated in international affairs, in part because of language barriers. At the same time, attendance at international meetings such as G8 summit meetings and United Nations meetings was considered quite special, and something to be proud of when the attendees came back to talk with their supporters. Attendance at international meetings represented an opportunity for them to be influenced by the ongoing dialogues among participants from other developed countries.

Hiroshi Ohta (2009) discussed Japan's actions on climate change from the point of view of diplomacy. He argued that Japan's domestic politics regarding climate change could be explained by five factors: political leadership, bureaucratic politics, environmental NGOs, the business sector, and public opinion polls. In his explanation of Japan's response to the ratification of the UNFCCC through the post-Kyoto negotiation process, he concluded that Japan mostly played the role of an "intermediate" or "support" state, and that this role was formulated by strong political leadership, political enthusiasm, and active participation of environmental NGOs. He reached this conclusion from the perspective of Robert Putnam's "two-level game" analysis (Putnam 1988). Ohta (2011) continued his analysis of Japan's climate change policy for the post-Kyoto period and concluded that a lack of strong and stable political leadership allowed relatively well-organized economic interests and METI to solidify their policy coalition in 2008–2009.

Kubo (2015) emphasized the ministerial controversy between the EA (later the MOE) and MITI (later METI), particularly in the late twentieth century. She argued that "changes occurred in Japan's political and administrative system from the second half of the 1990s through the 2000s," mainly for four reasons: changes in the government agency communities due to reorganization of the political arena, stronger activities of the environmental NGOs, reinforcement of cabinet functions through central government agency reforms, and changes in the nature of inter-ministerial coordination due to central government agency reforms.

Scientific community

Scientific experts can exert influence over Japan's decision-making in several ways. The Science Council of Japan has published several reports and recommendations regarding how Japan should respond to climate change (e.g., Science Council of Japan 2007, 2009). Scientific experts have also been invited to participate as members of various government committees to share their expertise on climate change. Nevertheless, it has been hard for scientists to be viewed as completely neutral in respect of the amount of emission reductions that should be made by Japan. Scientists who emphasize the urgent and serious nature of climate change can be perceived as "pro-environment," whereas those in support of continued economic prosperity as a precondition to emission reductions can be viewed as "anti-environment." Even the Science Council itself had to carefully consider the choice of members to draft the report. A similar thing can be said in the recommendation of experts as contributors to the IPCC reports.

Surprisingly, few academic studies are available on the role of scientists in Japan's domestic decision-making on climate change policy. Most of the literature includes scientists as one element in a more comprehensive view of decision-making process. Sato (2003) analyzed Japan's decision-making concerning ozone depletion and climate change by focusing on the power of knowledge, and found that scientific knowledge by itself did not affect decision-making in Japan; rather, it was a broader shift in attitudes, thinking, and perceptions of the Japanese people, which was not always due to science, that affected changes in Japan's decisions. Fisher (2003) conducted a series of interviews in Japan in 1999 concerning decision-making on climate change and concluded that the scientists working within Japanese academia on climate change were the same people who were leading the national labs to conduct research on climate change; in many cases they were also the leaders of governmental committees involved in determining national climate change policies.

Environmental NGOs and citizen groups

In studies of decision-making in many countries in the West, the role of environmental NGOs is considered crucial, but most studies of the subject in Japan tend to come to the conclusion that NGOs in Japan have very limited influence on the country's decisions on climate change mitigation policies. Traditionally, Japanese society did not support NGOs. The concept of NGOs was imported from Western nations and gradually diffused to Japan. In addition, the Japanese environmental NGOs working on climate change today are quite different from those of several decades ago.

Schreurs (2003) conducted a comparative study of environmental politics in Japan, Germany, and the United States on three atmospheric global environmental problems: acid rain, ozone depletion, and climate change. The author argued that Japan's environmental policy community, including environmental groups, had meager power in influencing the government's policy-making through the adoption of the Kyoto Protocol in 1997, but that the situation has changed since then and citizen groups have gained more influence. After interviewing nearly two dozen stakeholders in Japan in 1999, Fisher (2003) concluded that there was certainly room for Japanese civil society to play a role in the formulation of climate policy, suggesting that "Ecological Modernization" appeared to be occurring in Japan.

Industry and business

Japan's postwar economic growth is, in part, a result of strong government initiatives to support heavy industries after World War II. The government used to, and to some extent still does, intervene in markets to steer the economy in desired directions (Johnson 1982). Business and government are relatively well integrated unofficially, and they collaborate with each other in making and implementing policies. Thus, the relationship among Japanese industries, the business community, and METI has been quite strong compared with those among similar actors in other developed countries. Oshitani (2006) conducted a thorough comparison of the United Kingdom and Japan on the basis of theories related to interplay, mainly between governments and business groups, from 1988 to 1997. Japan and Britain were introduced as examples of consensus corporatism and majoritarian pluralism, respectively, and the author concluded that the styles of decision-making in the two countries influenced the development of climate change policies in ways quite different from each other.

The Keidanren (Japan Business Federation) is Japan's largest single economic and business federation, with a membership comprising 1,329 representative Japanese companies, 109 nationwide industrial associations, and 47 regional economic organizations as of June 2015. Its origin dates back to 1922, and it is the main representative of Japanese industries, particularly those in energy-intensive sectors. With close ties with the LDP and METI, the Keidanren has played a role as a shadow policymaker with regard to environmental issues (Iguchi et al. 2015; Kagawa-Fox 2012: 57). Another influential business group is Keizai Doyukai (the Japan Association of Corporate Executives), a private, nonprofit, nonpartisan organization that was founded in 1946. Its membership comprises approximately 1,300 top executives of about 940 corporations, with more of them being from less-energy-intensive businesses than is the case with the Keidanren. Its policy positions therefore sometimes differ from those of the Keidanren.

Watanabe (2011) conducted a comparative study between Germany and Japan on decision-making processes related to climate change. Compared with the situation in Germany, Japan's economic prosperity coalition – namely, the industry groups and MITI – was found to be more influential in determining Japan's climate change policies in the early stages of UNFCCC negotiations.

Social capital and local government

Japanese society has been characterized as being oriented towards cherishing harmony, group decision-making, and cultural homogeneity. Unlike in some other countries, Japanese culture has tended to prefer following others, rather than taking a leadership role. Hidefumi Imura (2005) argued that Japan was effective in the formation of informal institutions, comprising social norms or networks that supplement formal laws and institutions. These tendencies are reflected in the following chapters – for example, when we see that industry groups have advocated for a "voluntary approach" in major emission reduction policies in the industrial sector. They are also reflected in the approach local governments use in taking the initiative to set local agendas related to climate change mitigation.

There are no, if any, publications in English focusing on Japan's climate change policy-making and Japan's behavior in international negotiations under the UNFCCC over a relatively long timeframe (i.e., two to three decades). There is also no publication so far covering this topic in relation to the new international agreement, Paris Agreement, adopted at the 21st Conference of the Parties to the United Nations Framework Convention on Climate Change (COP21), held in Paris in December 2015. In this book, I cover all periods of Japanese climate change activities throughout the past three decades.

Structure of the book

The aim of this book is to examine Japan's decisions regarding responses to the climate change problem. The focus is particularly on high-level politicians, such as prime ministers and environment ministers, rather than on bureaucratic conflicts between ministries – the subject of many other studies. I examine how these politicians approached the climate change problem and how they perceived climate change among many other political, economic, and social issues that required their attention. Political leaders face many problems related to security, domestic politics, the economy, and social issues. For these leaders, climate change is merely one of many issues that need to be addressed, and incorporating climate change concerns with other more immediate issues such as employment, defense, economic development, and public health is key to finding proactive solutions (Rayner and Malone 1998).

The book covers a period of nearly three decades from the late 1980s to 2015. During these three decades, the countries' social, economic, and political fundamentals not related to climate change also changed in many ways. Care must be taken when analyzing Japan's behavior toward climate change in determining whether that behavior changed because the climate change problem changed or whether the countries' fundamental concerns changed. This book seeks to merge Japan's climate change policy-making with Japan's major domestic and foreign policy agendas of each era. Because this book makes a thorough evaluation of the history of this topic, it should be able to enlighten us on the future of Japan's climate change decision-making.

Chapter 2 covers the early days of the emergence of climate change as a global environmental problem that required a multilateral response. The chapter begins with a description of the post–World War II period when Japan began rebuilding, with most national efforts directed at rapid industrialization. Prioritization of industrial growth was effective in achieving economic growth and increased wealth, but it also simultaneously resulted in serious problems with local pollution and health hazards. In the late 1980s, heads of major economies started to demonstrate an eagerness to tackle global environmental problems such as ozone depletion, acid rain, desertification, and deforestation. Japanese politicians and the government were quick to respond to these issues, even though they might not have been fully aware of the significance of the problems. The UNFCCC was adopted in May 1992, and the Basic Environment Law was established in 1993.

Chapter 3 examines the period from COP1 to COP3, in which the Kyoto Protocol was negotiated and adopted, as well several years after COP3 to see how Japan followed up on relevant issues until Japan ratified the Kyoto Protocol in 2002. At COP1 in 1995, Japan announced its willingness to host COP3, to be held in Kyoto. After two years of intense negotiations, the Kyoto Protocol was adopted at COP3 in 1997. From one point of view, this could be seen as a success story for the Japanese government in terms of hosting COP3. In contrast, however, the adoption of the Kyoto Protocol was also the beginning of Japan experiencing the consequences of agreeing to a legally binding 6% emission reduction target from 2008 to 2012. The purpose of hosting COP3 and the consequences of it should be evaluated are discussed in the chapter.

Chapter 4 explains Japan's struggle to achieve a "post-2012" or "post-Kyoto" regime. The Kyoto Protocol entered into force in 2005, and Japan's emission reduction target also became internationally legally binding. Meanwhile, Japan began to insist on a post-Kyoto regime, in which all major GHG emitters would participate. Non-UN mechanisms, such as G8 negotiations and Asia-Pacific Partnership for Clean Development and Climate (APP), were utilized to stimulate political incentives to take action on climate change independent of what was to be agreed upon under the UNFCCC. Internally, the Japanese government initiated a series of intensive decision-making processes to set an emission reduction target for Japan for the year 2020. During this same time period, many people anticipated dramatic progress at the Copenhagen conference (COP15) in 2009, but it was a total failure in terms of developing a new multilateral institution in which all countries participate.

Chapter 5 begins with the March 11 earthquake in Tohoku and the ensuing serious nuclear accident at the Fukushima Daiichi Nuclear Power Plant. This incident completely changed Japan's energy policy. Because Japan's climate change mitigation policy was heavily dependent of the use of nuclear power, the roadmap towards low carbon development had to be reconsidered. The domestic climate change debate was more or less taken over by the debate on nuclear energy policy. An agreement known as the Durban Platform was reached at COP17, held in Durban, South Africa, in 2011. It aimed at arriving at a new agreement that would be applicable to all parties, including commitments for years beyond 2020, by 2015 at COP21. Japan's positions at the negotiations, its emission reduction targets for 2030, and its positions at the final stage of negotiations at COP21 are described in the chapter.

Chapter 6 summarizes how Japan has positioned itself in the global climate change debate and how it has responded to the problem. The chapter also makes brief remarks on the factors that have affected Japan's climate change policy-making. The chapter then goes on to forecast how Japan is likely to deal with climate change in the future.

References

Alber, Jens (1989) "Modernization, Cleavage Structures and the Rise of Green Parties and Lists in Europe," in Ferdinand Müller-Rommel ed., *New Politics in Western Europe: The Rise and Success of Green Parties and Alternative Lists*. Boulder, CO: Westview Press, 195–210.

Asuka-Zhang, Jusen (2003) "Development Assistance and Japan's Climate Change Diplomacy: Priorities and Future Options," in Paul G. Harris ed., *Global Warming and East Asia – The Domestic and International Politics of Climate Change*. London: Routledge, 152–166.

Betsill, Michele M. (2011) "International Climate Change Policy: Toward the Multilevel Governance of Global Warming," in Regina S. Axelrod, Stacy D. Vandeveer, and David Leonard Downie eds., *The Global Environment: Institutions, Law, and Policy*. Washington, DC: CQ Press, 111–131.

Betsill, Michele M. and Elisabeth Corell (2011) "NGO Influence in International Environmental Negotiations, A Framework of Analysis," *Global Environmental Politics*, 1(4), 65–85.

Boehmer-Christiansen, Sonja and Helmut Weidner (1995) *The Politics of Reducing Vehicle Emissions in Britain and Germany*. London: Pinter.

Bradley, Raymond S. (2011) *Global Warming and Political Intimidation*. Amherst, MA: University of Massachusetts Press.

Bryner, Gary (1993) *Blue Skies, Green Politics: The Clean Air Act of 1990*. Washington, DC: CQ Press.

Calder, Kent E. (1988) "Japanese Foreign Economic Policy Formation: Explaining the Reactive State," *World Politics*, 40(4), 517–541.

Chandler, William M. and Alan Siaroff (1986) "Postindustrial Politics in Germany and the Origins of the Greens," *Comparative Politics*, 18(3), 303–325.

Collier, Ute and Ragnar E. Löfstedt eds. (1997) *Cases in Climate Change Policy: Political Reality in the European Union*. London: Earthscan.

Collingridge, David and Colin Reeve (1986) *Science Speaks to Power: The Role of Experts in Policy Making*. New York: St. Martin's Press.

Conca, Ken and Ronnie D. Lipschutz (1993) "A Tale of Two Forests," in Ronnie D. Lipschutz and Ken Conca eds., *The State and Social Power in Global Environmental Politics*. New York: Columbia University Press, 1–18.

DeSombre, Elizabeth, R. (2000) *Domestic Sources of International Environmental Policy: Industry, Environmentalists, and U.S. Power*. Cambridge, MA: The MIT Press.

Doherty, Brian and Timothy Doyle (2006) "Beyond Borders: Transnational Politics, Social Movements and Modern Environmentalisms," *Environmental Politics*, 15(5), 697–712.

Downs, Anthony (1957) *An Economic Theory of Democracy*. New York: Harper & Row.

Dryzek, John S., David Downes, Christian Hunold, and David Schlosberg (2003) *Green States and Social Movements: Environmentalism in the United States, United Kingdom, Germany and Norway*. New York: Oxford University Press.

Ezrahi, Yaron (Spring 1980) "Utopian and Pragmatic Rationalism: The Political Context of Scientific Advice," *Minerva*, 18, 111–131.

Fire and Disaster Management Agency (FDMA) (2014) Shobo Hakusho [Fire and Disaster Management White Paper]. Tokyo: FDMA.

Fisher, Dana R. (2003) "Beyond Kyoto: The Formation of a Japanese Climate Change Regime," in Paul G. Harris ed., *Global Warming and East Asia – The Domestic and International Politics of Climate Change*. London: Routledge, 187–205.Frankental, Peter (2001) "Corporate Social Responsibility – A PR Invention?" *Corporate Communications: An International Journal*, 6(1), 18–23.

G8 Hokkaido Toyako Summit (2008) G8 Hokkaido Toyako Summit 2008 Leaders Declaration, 8 July 2008.

Gelbspan, Ross (2005) *Boiling Point*. New York: Basic Books.

Giddens, Anthony (2009) *Politics of Climate Change*. Cambridge: Polity Press.

Gieryn, Thomas F. (December 1983) "Boundary Work and the Demarcation of Science from Non-Science: Strains and Interests in Professional Ideologies of Scientists," *American Sociological Review*, 48, 781–795.

Gore, Al (1993) *Earth in Balance: Ecology and the Human Spirit*. New York: Plume Book.

Grubb, Michael (2014) *Planetary Economics, Energy, Climate Change and the Three Domains of Sustainable Development*. London: Earthscan, Routledge.

Guston, David (1999) "Stabilizing the Boundary between U.S. Politics and Science: The Role of the Office of Technology Transfer as a Boundary Organization," *Social Studies of Science*, 29(1), 87–111.Haas, Peter M. (1989) "Do Regimes Matter? Epistemic Communities and Mediteranean Pollution Control," *International Organizations*, 43(4), 377–404.

Haas, Peter M. (1992) "Introduction: Epistemic communities and International Policy Coordination," *International Organization*, 46(1), 1–35.

Haas, Peter M. (2007) "Epistemic Communities," in Daniel Bodansky, Jutta Brunnée, and Ellen Hey eds., *Oxford Handbook on International Environmental Law*. New York: Oxford University Press, 791–806.

Hajer, M.A. (1995) *The Politics of Environmental Discourse: Ecological Modernization and the Policy Process*. Oxford: Clarendon Press.

Harrison, Kathryn and Lisa McIntosh Sundstrom (2010) *Global Commons, Domestic Decisions: The Comparative Politics of Climate Change*. Cambridge: MIT Press.

Hattori, Takashi (2007) "The Rise of Japanese Climate Change Policy: Balancing the Norms Economic Growth, Energy Efficiency, International Contribution and Environmental Protection," in Mary E. Pettenger ed., *The Social Construction of Climate Change, Power, Knowledge, Norms, Discourses*. Aldershot: Ashgate, 75–97.

Heclo, Hugh (1978) "Issue Networks and the Executive Establishment," in Anthony King ed., *The New American Political System*. Washington, DC: American Enterprise Institute, 87–124.

Heinz, J., E. Laumann, R. Salisbury, and R. Nelson (1990) "Inner Circles or Hollow Cores," *Journal of Politics*, 52(2), 356–390.

Hochstetler, Kathryn (2012) "Democracy and the Environment in Latin America and Eastern Europe," in Paul F. Steinberg and Stacy D. VanDeveer eds., *Comparative Environmental Politics*. Cambridge, MA: The MIT Press, 199–229.

Hook, Glenn D., Julie Gibson, Christopher W. Hughes, and Hugo Dobson (2012) *Japan's International Relations: Politics, Economics and Security*. Abingdon: Routledge.

Howlett, M. and M. Ramesh (2003) *Studying Public Policy: Policy Cycles and Policy Subsystems*. Ontario: Oxford University Press.

Iguchi, Masahiko, Alexandru Luta and Steinar Andresen (2015) "Japan's Climate Policy: Post-Fukushima and Beyond," in Guri Bang, Arild Underdal and Steinar Andresen eds., *The Domestic Politics of Global Climate Change: Key Actors in International Climate Cooperation*. Cheltenham: Edward Elgar Publishing.

Imura, Hidefumi (2005) "Evaluating Japan's Environmental Policy Performance," in Hidefumi Imura and Miranda Schreurs eds., *Environmental Policy in Japan*. Cheltenham, UK: The World Bank and Edward Elgar, 342–359.

Inglehart, Ronald (1982) *Changing Values and the Rise of Environmentalism in Western Societies*. Berlin: International Institute for Environment and Society.

Inoguchi, Takashi (1991) *Japan's International Relations*. London: Pinter.

Inoguchi, Takashi (2012) *Japan's Foreign Policy in an Era of Global Change*, Bloomsbury Academic Collections. London: Bloomsbury Academics.

Intergovernmental Panel on Climate Change (IPCC) (2014) *Climate Change 2014: Mitigation of Climate Change*, Contribution of Working Group III to the Fifth Assessment Report of the Intergovernmental Panel on Climate Change. Geneva: IPCC.

Intergovernmental Panel on Climate Change (IPCC) (2015) *Climate Change 2014: Synthesis Report*, Contribution of Working Groups I, II and III to the Fifth Assessment Report of the Intergovernmental Panel on Climate Change. Geneva: IPCC.

International Energy Agency (IEA) (2015) *Energy and Climate Change, World Energy Outlook Special Report*. Paris: IEA.

Jacques, Peter J. (2012) "A General Theory of Climate Denial," *Global Environmental Politics*, 12(2), 9–17.

Jancar-Webster, Barbara (1998) "Environmental Movement and Social Changes in the Transition Countries," *Environmental Politics*, 7(1), 69–90.

Jänicke, Martin and Helmut Weidner (1995) *Successful Environmental Policy: A Critical Evaluation of 24 Cases*. Berlin: Edition sigma.

Jasanoff, Sheila (1990) *The Fifth Branch, Science Advisers as Policymakers*. Cambridge: Harvard University Press.

Jasanoff, Sheila (2005) *Designs of Nature: Science and Democracy in Europe and the United States*. Princeton, NJ: Princeton University Press.

Johnson, Chalmer (1982) *MITI and the Japanese Miracle, 1925–1975*. Stanford: Stanford University Press.

Kagawa-Fox, Midori (2012) *The Ethics of Japan's Global Environmental Policy*. Abingdon: Routledge.

Kameyama, Yasuko (2003) "Climate Change as Japanese Foreign Policy: From Reactive to Proactive," in Paul G. Harris ed., *Global Warming and East Asia – The Domestic and International Politics of Climate Change*. London: Routledge, 135–151.

Kawashima, Yasuko (2000) "Japan's Decision-Making about Climate Change Problems: Comparative Study of Decisions in 1990 and in 1997," *Environmental Economics and Policy Studies*, 3(1), 29–57.

Kawashima, Yasuko (2001) "Japan and Climate Change: Responses and Explanations," *Energy and Environment*, 12(2&3), 167–180.

Keck, Margaret E. and Kathryn Sikkink (1998) *Activists Beyond Borders: Advocacy Networks in International Politics*. Ithaca, NY: Cornell University Press.

Kingdon, J. (1995) *Agendas, Alternatives, and Public Policies*, 2nd edition. New York: Addision-Wesley.

Kitschelt, Herbert P. (1993) "The Green Phenomenon in Western Party Systems," in Sheldon Kamieniecki ed., *Environmental Politics in the International Arena: Movements, Parties, Organizations and Policy*. Albany: SUNY Press, 93–112.

Knoke, David, Franz Urban Pappi, Jeffery Broadbent, and Yutaka Tsujinaka (1996) *Comparing Policy Networks: Labor Politics in the U.S., Germany, and Japan*. Cambridge: Cambridge University Press.

Kolk, Ans and D. Levy (2001) "Winds of Change: Corporate Strategy, Climate Change and Oil Multinationals," *European Management Journal*, 19(5), 501–509.

Kolk, Ans and J. Pinkse (2005) "Business Responses to Climate Change: Identifying Emergent Strategies," *California Management Review*, 47(3), 6–20.

Kraft, Michael E. and Sheldon Kamieniecki (2007) "Analyzing the Role of Business in Environmental Policy," in Michael E. Kraft and Sheldon Kamieniecki eds., *Business and Environmental Policy: Corporate Interests in the American Political System*. Cambridge, MA: MIT Press, 3–32.

Kubo, Haruka (2015) "The Possibilities for Climate Change Policy Integration as Seen from Japan's Political and Adminisrative System," in Hidenori Niizawa and Toru Morotomi eds., *Governing Low-Carbon Development and the Economy*. Tokyo: United Nations University Press, 185–206.

Latour, Bruno (1987) *Science in Action: How to Follow Scientists and Engineers through Society*. Cambridge, MA: Harvard University Press.

Lee, Yok-Shiu F. and Alvin Y. So eds. (1999) *Asia's Environmental Movements: Comparative Perspectives*. Armonk, NY: M.E. Sharpe.

Lijphart, A. (1999) *Patterns of Democracy: Government Forms and Performance in Thirty-Six Countries*. New Haven, CT: Yale University Press.

Litfin, Karen (1994) *Ozone Discourses, Science and Politics in Global Environmental Cooperation*. New York: Columbia Press.

Marsh, D. and R.A.W. Rhodes (1992) *Policy Networks in British Government*. Oxford: Oxford University Press.

Meadowcroft, James (2004) "Deliberative Democracy," in Robert F. Durant, Daniel J. Fiorino, and Rosemary O'Leary eds., *Environmental Govenance Reconsidered: Challenges, Choices, and Opportunities*. Cambridge, MA: The MIT Press, 183–217.

Meteorological Agency (2015) *Ijokisho Repoto* [Extreme Weather Report]. Tokyo: Meteorological Agency.

Miller, Clark A. (2007) "Democratization, International Knowledge Institutions, and Global Governance," *Governance*, 20(2), 325–357.

Miller, Clark A. and Paul N. Edwards (2001) "Introduction: The Globalization of Climate Science and Climate Politics," in C.A. Miller and P.N. Edwards eds., *Changing Atmosphere: Expert Knowledge and Environmental Governance*. Cambridge, MA: The MIT Press, 1–30.

Mills, C. Wright (1956) *The Power Elite*. Oxford: Oxford University Press.

Muramatsu, Michio and Yutaka Tsujinaka (1992) *Nihon no Seiji* [Japanese Politics]. Tokyo: Yuhikaku.

Nester, William (1990) *The Foundation of Japanese Power: Continuities, Changes, Challenges*. New York: M.E. Sharpe.

Nye, Joseph (2004) *Soft Power: The Means to Success in World Politics*. New York: Public Affairs.

Ohta, Hiroshi (2009) "Japanese Foreign Policy on Climate Change: Diplomacy and Domestic Politics," in Paul G. Harris ed., *Climate Change and Foreign Policy*. Abingdon: Routledge, 36–52.

Ohta, Hiroshi (2011) "Japanese Climate Change Policy: Moving Beyond the Kyoto Protocol," in Hans Günter Brauch, Úrsula Oswald Spring, Czeslaw Mesjasz, John Grin, Patricial Kameri-Mbote, Béchir Chourou, Pál Dunay, and Jörn Birkmann eds., *Coping with Global Environmental Change, Disasters and Security: Threats, Challenges, Vulnerabilities and Risks*, 4. Berlin: Springer, 1381–1391.

O'Neill, Michael (2012) "Political Parties and the 'Meaning of Greening' in European Politics," in Paul F. Steinberg and Stacy D. VanDeveer eds., *Comparative Environmental Politics, Theory, Practice and Prospects*. Cambridge, MA: The MIT Press, 115–142.

Oppschoor, H. and J. van der Straaten (1993) "Sustainable Development: An Institutional Approach," *Ecological Economics*, 7, 203–222.

O'Riordan, Tim R. and Jill Jäger eds. (1996) *Politics of Climate Change: A European Perspective*. London: Routledge.

Oshitani, Shizuka (2006) *Global Warming Policy in Japan and Britain*. Macmillan: Manchester University Press.

Parson, Edward, A. (2003) *Protecting the Ozone Layer*. Oxford: Oxford University Press.

Pempel, T.J. (1979) "Japanese Foreign Economic Policy: The Domestic Bases for International Behavior," *International Organization*, 31(4), 723–774.

Pempel, T.J. (1992) "Japanese Democracy and Political Culture: A Comparative Perspective," *Political Science and Politics*, 25, 5–12.

Pinguelli-Rosa, Luiz and Mohan Munasinghe (2002) *Ethics, Equity and International Negotiations on Climate Change*. Cheltenham: Edward Elgar.

Princen, Thomas and Matthias Finger (1994) *Environmental NGOs in World Politics: Linking the Local and the Global*. London: Routledge.

Putnam, Robert (1988) "Diplomacy and Domestic Politics: The Logic of Two-Level Games," *International Organization*, 42(3), 427–460.

Rayner, Steve and Elizabeth L. Malone (1998) "Ten Suggestions for Policy Makers," in Steve Rayner and Elizabeth L. Malone eds., *Human Choice & Climate Change: Vol. 4 What Have We Learned?* Columbus, OH: Battelle Press, 109–138.

Rhodes, R.A.W. (1990) "Policy Networks," *Journal of Theoretical Politics*, 2(3), 293–317.

Richardson, J. and A. Jordan (1979) *Governing under Pressure: British Democracy in a Post-Parliamentary Democracy*. Oxford: Martin Robertson.

Ringius, Lasse, Asbjørn Torvanger, and Arild Underdal (2002) "Burden Sharing and Fairness Principles in International Climate Policy," *International Environmental Agreements: Politics, Law and Economics*, 2, 1–22.

Rüdig, Wolfgang (1985) "The Greens in Europe: Ecological Parties and the European Elections of 1984," *Parliamentary Affairs*, 38(1), 56–72.

Sabatier, P. and H. Jenkins-Smith eds. (1993) *Policy Change and Learning: An Advocacy Coalition Approach*. Boulder, CO: Westview Press.

Sagar, Ambuj (2000) "Wealth, Responsibility and Equity: Exploring an Allocation Framework for Global GHG Emissions," *Climate Change*, 45, 511–527.

Sands, H. Peter (1992) *Lessons Learned in Global Environmental Governance*. Washington, DC: World Resources Institute.

Sato, Atsuko (2003) "Knowledge in the Global Atmospheric Policy Process: The Case of Japan," in Paul G. Harris ed., *Global Warming and East Asia – The Domestic and International Politics of Climate Change*. London: Routledge, 167–186.

Schmidheiny, Stephan with BCSD (Business Council for Sustainable Development) (1992) *Changing Course: A Global Business Perspective on Development and the Environment.* Cambridge, MA: The MIT Press.

Schreurs, Miranda A. (2003) *Environmental Politics in Japan, Germany, and the United States.* Cambridge: Cambridge University Press.

Science Council of Japan (2007) *Chikyu Ondanka to Enerugi* [Global Warming and Energy], Report 22 March 2007.

Science Council of Japan (2009) *Chikyu Ondanka Mondai Kaiketsu no Tameni* [Solution to the Global Warming Problem], Report 10 March 2009.

Shackley, Simon and Brian Wynne (August 1995) "Global Climate Change: The Mutual Construction of an Emergent Science-Policy Domain," *Science and Public Policy*, 22, 218–230.

Shukla, P.R. (1999) "Justice, Equity and Efficiency in Climate Change, A Developing Country Perspective," in F.L. Toth ed., *Fair Weather? Equity Concerns in Climate Change.* London: Earthscan, 150–155.

Skjærseth, Jon Birget and Tora Skodvin (2003) *Climate Change and the Oil Industry: Common Problem, Varying Strategies.* Manchester: Manchester University Press.

Smith, M.J. (1993) *Pressure, Power and Policy: State Autonomy and Policy Networks in Britain and the United States.* Hemel Hempstead: Harvester Wheatsheaf.

Social Learning Group (2001) *Learning to Manage Global Environmental Risks. Vol. 1: A Comparative History of Social Responses to Climate Change, Ozone Depletion, and Acid Rain.* Cambridge, MA: The MIT Press.

Sprinz, Detlef F. (2009) "Long-Term Environmental Policy: Definition, Knowledge, Future Research," *Global Environmental Politics*, 9(3), 1–8.

Sprinz, Detlef F. and T. Vaatoranta (1994) "The Interest-Based Explanation of International Environmental Policy," *International Organization*, 48(1), 77–105.Steinberg, Paul F. and Stacy D. VanDeveer eds. (2012) *Comparative Environmental Politics.* Cambridge, MA: The MIT Press.

Stern, Nicholas (2013) *Ethics, Equity and the Economic of Climate Change – Paper 2: Economic and Politics*, Center for Climate Change Economic and Policy Working Paper No. 97b. London: London School of Economics and Political Science.

Tarrow, Sidney (1998) *Power in Movement: Social Movements and Contentious Politics*, 2nd edition. Cambridge, UK: Cambridge University Press.

Tiberghien, Yves and Miranda A. Schreurs (2010) "High Noon in Japan: Embedded Symbolism and Post-2001 Kyoto Protocol Politics," *Global Environmental Politics*, 7(4), 70–91.

Tóth, Ferenc L. (1999) "Fair Concerns in Climate Change," in F.L. Tóth ed., *Fair Weather? Equity Concerns in Climate Change.* London: Earthscan, 1–10.

Underdal, Arild (1997) "Leadership in International Environmental Negotiations: Designing Feasible Solutions," in A. Underdal ed., *The Politics of International Environmental Management.* Dordrecht: Kluwer Academic Publishers, 101–127.

UNFCCC (2010) Decision 1/CP.16 The Cancun Agreements: Outcome of the Work of the Ad Hoc Working Group on Long-Term Cooperative Action under the Convention, FCCC/CP/2010/7/Add.1.

United Nations Framework Convention on Climate Change (UNFCCC) (2015) Paris Agreement, FCCC/CP/2015/L.9.

Urpelainen, Johannes (2012) "Global Warming, Irreversibility, and Uncertainty: A Political Analysis," *Global Environmental Politics*, 12(4), 68–85.

Vogel, D. (1986) *National Style of Regulation: Environmental Policy in Great Britain and the United States*. Ithaca, NY: Cornell University Press.

Wapner, Paul K. (1996) *Environmental Activism and World Civil Politics*. Albany: SUNY Press.

Watanabe, Rie (2011) *Climate Policy Changes in Germany and Japan*. Abingdon: Routledge.

Weinberg, A.M. (1972) "Science and Trans-Science," *Minerva: A Review of Science, Learning, and Policy*, 10(2), 209–222.

Weingart, Peter (1982) "The Scientific Power Elite – A Chimera," in Norbert Elias, Herminio Martins, and Richard Whitley eds., *Scientific Establishments and Hierarchies*. Dordrecht: Reidel, 71–88.

White House (2001) *Press Release: Text of A Letter from President George W. Bush to Senators Hagel, Helms, Craig, and Roberts*. Washington, DC, March 13. Available online at: http://georgewbush-whitehouse.archives.gov/news/releases/2001/03/20010314.html (accessed 30 January 2016).

Young, Oran R. (1994) *International Governance: Protecting the Environment in a Stateless Society*. Ithaca, NY: Cornell University Press.

Zehr, Stephen (December 2005) "Comparative Boundary Work: US Acid Rain and Global Climate Change Policy Deliberations," *Science and Public Policy*, 32, 445–456.

2 Emergence of the climate change problem and adoption of the UNFCCC (1980s–1994)

Overview

Furuikeya kawazu tobikomu mizu no oto
(An old pond / a frog jumps in / sound of water)

Basho Matsuo[1]

This chapter describes the early history of climate change policy in Japan. In many ways, Japan's contemporary institutions emerged after the end of World War II in 1945. Rapid industrialization throughout the 1950s and 1960s brought about economic prosperity, but it also led to serious environmental pollution in local communities.

In the late 1980s, officials in Japan noticed that climate change could become an important global political agenda. Japanese politicians and the government jumped into the emerging "pond" of the global warming debate without much hesitation and, presumably, without being fully aware of the significance of the problem. Similar to an image of Basho Matsuo's haiku in which the sound of water was highlighted in total tranquility, Japan's jump into the pond was accompanied by a splash that rippled throughout the country, opening people's eyes about global environmental problems. The United Nations Framework Convention on Climate Change (UNFCCC) was adopted in May 1992, followed by the United Nations Conference of the Environment and Development (UNCED) held in Rio de Janeiro in June of the same year. Japan ratified the UNFCCC on 28 May 1993.

The early development of Japan's environmental policy

Japan's major experience with industrial pollution dates back to the 1950s when it was in the midst of economic recovery after World War II. Both political leaders and bureaucratic officials had prioritized economic growth by rapid industrialization, with little awareness of the serious consequences in terms of local pollution and human health (Wilkening 2004). Toxic effluent discharged into the atmosphere, rivers, and sea altered various environmental fundamentals upon which the Japanese people used to depend in their daily lives. Minamata disease, Niigata Minamata disease, Itai-itai disease, and Yokkaichi asthma, known as the "Four Big Pollution Diseases," were the best-known local health hazards resulting from toxic chemical discharge into the surrounding environment (Broadbent 1998; Committee on Japan's Experience in the Battle against Air Pollution 1997; Matsuno 2007; OECD 1977; Walker and Cronon 2011). Basic Law for Pollution Control was adopted in 1967 to respond to these imminent incidents.

The 64th Annual Meeting of the Japanese Diet in 1970 was called the "Pollution Diet," because it passed 15 pieces of legislation concerning pollution. The Environment Agency (EA) was established in 1971 to deal with local pollution. Seven typical types of pollution – air pollution, water pollution, soil pollution, noise, vibration due to industrial activity and heavy traffic, leveling down of land surface as a result of the tremendously large amounts of underground water extracted, and odor – were designated as the main scope of the EA's authority to regulate industrial activities. The administrative authority of the EA also covered protection and conservation of nature in Japan. National parks were established in 1934 to conserve scenic beauty, but it was only after the establishment of the EA that protection of wildlife was added as part of the national parks' conservation mandate. The Nature and Environment Conservation Law was enacted in 1972.

The late 1960s and early 1970s also marked the beginning of the global environmental movement. Many Western publications of the time projected a doomsday scenario for human beings because of the rapid population growth in the South, depletion of natural resources (especially oil), and heavy contamination in many parts of the world (e.g., Ehrlich 1968; Hardin 1968). The Club of Rome published the report "The Limits to Growth," utilizing a computer simulation of exponential population and economic growth coupled with a limited supply of natural resources (Meadows et al. 1972). The United Nations Conference on the Human Environment was held in Stockholm in 1972 to discuss ways to deal with rapidly growing populations and limited natural resources and food (Tolba et al. 1992). At the time, Japan was situated somewhere between the developed and the developing countries in terms of economic growth, and it was unable to become fully involved in the debate.

Most of the pollutants in the 1950s and 1960s were discharged by industrial complexes, so regulatory measures targeted those firms and locations to reduce the amount of discharge. The oil crises of 1973 and 1979 also motivated industries to invest in energy-saving technologies. The amount of investment in pollution prevention technology jumped drastically during this period, reaching 96.45 billion yen by 1975 and accounting for 17.7% of total capital investment (Miyamoto 2013). The Law on the Rational Use of Energy, enacted in 1979, was considered to be particularly effective in motivating industries to improve energy efficiency. Through these combined efforts, the level of local pollution had decreased dramatically by the end of the 1970s (EA 1983).

Japan has depended on overseas supplies for most of its energy resources. Therefore, energy security has long been a priority; in practice, this has meant maintaining a balanced mix of various types of energy to assure a stable supply. In the 1960s, the government began to show a strong willingness to build nuclear power plants in Japan to reduce dependence on fossil fuels. Even though Japan was the victim of atomic bombs during World War II, the Japanese government did not hesitate to allocate a huge amount of the national budget to technology development for nuclear power plants in the 1960s. Investment in nuclear power plants in Japan was also supported by the United States. The two oil crises in the 1970s further increased support for nuclear power. The amount of nuclear power supplied grew from 0.6% of the total primary energy supply in 1973 to 9.4% in 1990, and it continued to grow thereafter.

In general the Japanese people's interest in the global environment began to emerge in the 1980s after Japan's economic development and people's lifestyles caught up with those in the average Western (i.e., industrialized) nation. Former United States president Jimmy Carter's report, titled "The Global 2000 Report to the President" (Barney 1980), was published in 1980 and received a great deal of interest in Japan, but it did not have much influence on the government's decisions. Focus on environmental governance at that time was still mostly on pollution at the national and local levels. Japanese policymakers believed that Japan was still in the midst of industrialization, so they were positioning themselves more from a newly emerging economy's viewpoint. Therefore, Japan, along with many other countries at the time, considered that pollution abatement policies (i.e., environmental policies) were to be implemented at the national level only.

Japan discovers climate change

Apart from many policy makers' perceptions that Japan was still at a developing stage, Japan had become one of the world's major economies by the early 1980s. Environmental quality had also improved dramatically, so people generally became less concerned about the environmental problems at the local scale (Prime Minister's Office 1984). Nevertheless, two new types of environmental issues were gradually becoming apparent.

The first new issue came about as a consequence of people's higher standards of living. As people became wealthier and enjoyed a more luxurious lifestyle, more of them began to own cars and energy-consuming home appliances such as larger refrigerators and air conditioners. Consumer goods were sold with extensive packaging materials, which were discarded almost immediately after the goods were purchased. These new lifestyle conditions created environmental challenges, including the need for sound waste management and efficient use of raw materials, along with the degradation of air quality as a result of heavy traffic congestion. In this new scenario, the source of pollutants was not only industry but also households, so new environmental policies were needed to target not only industrial activities but the entire economy.

The second issue was the emergence of what was called "global environmental problems." This type of problem included acid rain, destruction of the ozone layer, climate change, desertification, and loss of wildlife and biological diversity. These problems could not be solved by a single nation, and Japan, as one of the major economic superpowers of the time, was expected by other countries to respond to these global issues. The issue of ozone-depleting substances did not attract much attention in Japan, perhaps because skin cancer was little known among Japanese people; this may have made it more difficult for people to be sensitive to the problem.

In addition, Japanese companies that had invested in developing countries in Southeast Asia and other areas were criticized by various environmental non-governmental organizations (NGOs) for destroying local environments in these countries (Dauvergne 1997). At the time, given the relatively strong yen in the currency market, it was cheaper for Japanese companies to invest in developing countries – particularly those in Southeast Asia – to manufacture consumer goods abroad and import the products to Japan, changing the production process even of large Japanese corporations (Tabb 1995: 257–258). In a way, this can be seen as exporting the production-related environmental hazards to the developing countries because, for example, Japan's forest coverage rate had been maintained as imports of wood products from Asian countries increased. Degradation of the natural environment in developing countries by Japanese firms was perceived by the Japanese government as a global environmental issue that needed to be mitigated, even if only from the perspective of diplomacy (Ueta 2005).

Few Japanese were aware of the terms "global warming" or "climate change" until the late 1980s. The origins of the theory of climate change hail back to the nineteenth century. Svante Arrhenius, who is famous as the pioneer of climate change research, pointed out that CO_2 in the atmosphere absorbed heat released from Earth's surface and reflected it back down to Earth (Arrhenius 1896). At the time, people were not unduly worried because the prevailing view was that the level of CO_2 released from burning fossil fuels was not sufficient to become a global problem. Kenji Miyazawa, a well-known poet and writer of fiction for children in the early 1900s, wrote a story about a farmer named Budori and his village who were suffering from an extremely cold winter; all of their crops were damaged by ice-cold climate (Miyazawa 1932). Because the people were starving, Budori decided to take action. He heard that a volcanic eruption would emit CO_2, which would lead to warming. He decided to cause an eruption himself, and the village was saved. It is uncertain how Miyazawa gained his knowledge relating to the science of global warming, but, in any case, the mechanism of global warming caused by CO_2 in the atmosphere was understood by most Japanese people reading the story, even though most people were unaware that such a thing could actually occur. Meanwhile, scientists in the West continued investigating the mechanism and severity of climate change. Measurements of CO_2 concentrations in the atmosphere were taken in the 1950s at Mt. Mauna Loa in Hawaii. These measurements confirmed suspicions: CO_2 concentrations were indeed rising, and more and more scientists began to focus on this problem over time.

Almost no action to study climate change mechanisms was taken by Japanese scientific communities in the 1970s. Among the very few who did take action was Shukuro Manabe, who moved to the United States to work for the National Weather Service and National Oceanic and Atmospheric Administration, but his research activities were influenced by American academic societies rather than by those in Japan. Climate change induced by anthropogenic greenhouse gas (GHG) emissions was a part of a discussion during the Kisho Gakkai (Japan Society of Meteorology) annual meeting in 1978, but this did not develop into wider discussions within Japanese academia.

In part because of criticism from abroad, as well as Japan's increasing willingness to make an "international contribution" in the international arena, to improve its international reputation the Japanese government gradually started taking positive action toward environmentally sound policies. This international contribution was considered to be the key to solidifying Japan's status as an accepted member of the world community (Busby 2010: 95). One of Japan's first attempts to engage in multilateral action toward addressing global environmental problems was with the establishment of the World Commission on Environment and Development (WCED, or the so-called Brundtland Commission) set up in 1984 under the UN General Assembly. The Japanese government proposed the establishment of the commission and made a substantial financial contribution, although this fact is not widely acknowledged. The final report in 1987, "Our Common Future," influenced various global actions related to the environment and development, including the creation of UNCED in 1992 (WCED 1987).

At the international level, researchers gradually started to press policymakers to act on climate change. Scientists and policymakers convened meetings in Villach, Austria, in 1980, 1983, 1985, and 1987, and in Bellagio, Italy, in 1987 (Jäger and O'Riordan 1996: 12–21). The meetings were attended mostly by experts and policymakers from the West: at the time, there was no one in Japan who was able to translate these climate change dialogues into policy change. These meetings bore fruit in June 1988, when the World Conference on the Changing Atmosphere was held in Toronto, Canada. The Conference called on all developed countries to cut their CO_2 emissions by 20% from 1987 levels by 2005.

The Group of Seven meeting (G7 Summit) was also held in Toronto just before the Conference on the Changing Atmosphere. At the Toronto G7 Summit, participants discussed not only political and economic issues but also growing concerns about global environmental degradation. Particular attention was paid to problems related to the atmosphere, such as acid rain, ozone depletion, and climate change. Many of the participants at this meeting, including Prime Minister Margaret Thatcher of the United Kingdom, Chancellor Helmut Kohl of West Germany, and President François Mitterand of France, were eager to discuss climate change. Japanese Prime Minister Noboru Takeshita attended the meeting and discovered, to his surprise, that other participants were discussing climate change. On the closing day of the G7 meeting, climate change was embedded in the "Economic Declaration," meaning that climate change, along with other global environmental problems, was acknowledged to have economic implications.

The Declaration stated:

> Further action is needed. Global climate change, air, sea and fresh water pollution, acid rain, hazardous substances, deforestation, and endangered species require priority attention. It is, therefore, timely that negotiations on a protocol on emissions of nitrogen oxides within the framework of the Geneva Convention on Long-range Transboundary Air Pollution be pursued energetically. The efforts of the United Nations Environment Programme (UNEP) for an agreement on the trans-frontier shipment of hazardous wastes should also

be encouraged as well as the establishment of an inter-governmental panel on global climate change under the auspices of UNEP and the World Meteorological Organization (WMO). We also recognize the potential impact of agriculture on the environment, whether negative through over-intensive use of resources or positive in preventing desertification.

(Group of Seven 1988)

Upon returning from the G7 meeting, Takeshita instructed officials in the Japanese government to prepare for a climate change debate. It was then that many Japanese policymakers realized that climate change was not merely a scientific debate, but that it was also already in the international political arena.

A huge group of scientists became part of a formal institution, the Intergovernmental Panel on Climate Change (IPCC), established by UNEP and WMO in November 1988. The objective was to provide the world with a clear scientific view on the latest knowledge about climate change and its potential environmental and socio-economic impacts. Because most of the scientific studies of climate change had been produced by scientists in the West (i.e., the United States, the United Kingdom, and European countries), only a few Japanese scientists participated in drafting the first assessment report, published in 1990.

Japanese policymakers sensed that climate change was going to be a controversial problem and had to respond to it quickly. The Council of Relevant Ministers on Global Environmental Conservation was established by the government in May 1989. The Council dealt with global environmental issues in general, but climate change was considered to be the most urgent, particularly because it has significant impacts on the Japanese economy because it was directly connected to energy use. The EA also created a new bureau, the Global Environment Protection Office, inside the agency in the same year. The Office was promoted to a Division the following year.

Japan's first attempt to set a GHG emission target

Although Japan was ready and willing to make a positive contribution to climate change mitigation at this time, it was also cautious about setting numerical targets for CO_2 emissions. The Noordwijk Ministerial Conference in November 1989 was the first ministerial meeting on climate change attended by environmental ministers from approximately 70 countries (Brenton 1994: 171). At the conference, a number of countries, including the Netherlands and Sweden, argued that numerical targets for CO_2 emissions should be set. The United States opposed this position because of remaining scientific uncertainty. China and the Soviet Union also opposed the idea of setting emission reduction targets for themselves, arguing that industrialized countries should take the lead in emissions mitigation. Japan, which objected to setting a flat-rate reduction target, explained that a flat-rate reduction would be unfair to those countries that had already achieved relatively high levels of energy efficiency. Setting the same reduction rate, such as "X percent by 19XX" for all countries, would put Japan at a disadvantage. The statement that was finally agreed upon at the conference stated that many (but not all) industrialized countries recognized that stabilization of CO_2 emissions at current levels should be achieved by 2000 at the latest.

Japan's opposition to emission reduction targets at the Noordwijk Conference was front-page material in Japanese newspapers. Members of the public who were becoming more aware of global environmental problems criticized the government's negative response at the meeting. Some Diet members claimed that Japan should take the lead in tackling global environmental problems (Ohta 2000: 104). The government officials who attended the Noordwijk Conference were pressured by the Diet members to reconsider their position.

In 1990, after the Netherlands, Germany, and the United Kingdom announced their own national goals, the Japanese government started working on the determination of a national emission target to reduce CO_2 emissions. Germany announced a "25% reduction of CO_2 emission from the 1987 level by 2005," whereas the United Kingdom committed to "stabilization at the 1990 level by 2005." Seeing the behavior of these leading countries, Japan hastened to do the same. In June, the Council of Relevant Ministers agreed that Japan would set its CO_2 emission target for 2000.

Discussions on the target took place exclusively inside the central government, mainly among the EA, the Ministry of International Trade and Industry (MITI), and the Agency for Natural Resources and Energy (ANRE), which was an administrative office under MITI. Although the Japanese people were more informed about climate change than they had been in previous years, there was no pressure from environmental NGOs to set stringent targets for emission reductions, mainly because there were no influential environmental NGOs working on climate change at that time in Japan. The three organizations discussed the target during the summer of 1990, but their views differed. The EA insisted that emission stabilization by 2000 at 1990 levels was necessary to mitigate climate change. NRE and MITI noted that energy consumption in Japan was expected to increase in the next decade under the precondition that the past decade's growth in gross domestic product (GDP) would be repeated in the next decade; thus. stabilization of CO_2 emissions would be impossible, even with rapid development and diffusion of a variety of innovative energy-related technologies. Their views were based on the report "New Earth 21: Action Program for the Twenty-First Century," which involved the accelerated introduction of "clean energy sources" such as nuclear power plants, introduction of new and renewable energy sources in the future, and the deployment and diffusion of environmentally sound technologies (RITE 1990). The plan did not assume stabilization of CO_2 emissions, at least not in the near term.

Other ministries such as the Ministry of Foreign Affairs (MOFA) and Ministry of Agriculture, Forestry and Fisheries (MAFF) supported the EA's position. The gap between the two positions was mediated by members of the Liberal Democratic Party (LDP) (Takemoto 1991).The final agreement took a middle ground approach by using the concept of "stabilization of emissions per capita."

At that time, the Japanese population was expected to grow at a rate of about 6% from 1990 to 2000. Thus, per capita stabilization meant that emissions would be allowed to increase by 6% (Schreurs 1996).

The Action Program to Arrest Global Warming (Global Warming Action Plan) was adopted in October 1990 (see Box 2.1). A two-tier target was set for CO_2 emissions. The first target was that CO_2 emissions on a per capita basis in 2000 and beyond be stabilized at approximately the same level as in 1990. The target was strengthened in the second target, which was stabilization of absolute emissions by 2000 and beyond at the 1990 level on the condition that innovative technology would be created in 1990–2000. The first tier was a reflection of the compromise among the ministries, and the second tier was added to reflect the EA's willingness to set a more ambitious target. The Action Plan also briefly touched upon other GHGs by stating that methane emissions as well as other GHG emissions should also not grow from current levels. No quantitative target was set for sequestration of carbon by sinks such as forests, but the plan stated that measures should be taken for forest management in Japan and that Japan should also take measures for the conservation of forests at the global level.

Box 2.1   The Action Program to Arrest Global Warming (Government of Japan 1990) (excerpt) 23 October 1990

III. Targets Under the Action Program

The targets for the limitation of GHGs emission shall be set as follows:

(1) The Government of Japan, based on the common efforts of the major industrialized countries to limit CO_2 emissions, establishes the following targets for the stabilization of Japan's CO_2 emission.

    a   The emissions of CO_2 should be stabilized, on a per capita basis in the year 2000 and beyond at about the same level as in 1990, by steadily implementing a wide range of measures under this Action Program, as they become feasible, through the utmost efforts by both the government and private sectors.

    b   Efforts should also be made, along with the measures above, to stabilize the total amount of CO_2 emission in the year 2000 and beyond at about the same level as in 1990, through progress in the development of innovative technologies, etc., including those related to solar, hydrogen and other new energies, as well as fixation of CO_2 at a pace and scale greater than currently predicted.

(2) The emissions of methane gas should not exceed the present level. To the extent possible, the emissions of nitrous oxide and other GHGs should not be increased.

With respect to sinks of CO_2, efforts should be made to work for the conservation and development of forests, greenery in urban areas and so forth in Japan and also to take steps to conserve and expand forests on a global scale, among others.

Debates on climate change among both the bureaucrats and the policymakers, along with the establishment of IPCC in 1988, stimulated the scientific community in Japan. Traditionally, Japan used an accumulation of expert knowledge to project the short-term weather patterns used for daily weather forecasts. The most serious factor that hampered the development of climate-related science was the slow development of computers. Only in 1980 did the Meteorological Research Institute install a computer designed for the development of atmospheric circulation models, and scientific investigation (including of the global warming potential of GHGs) began. In 1990, the National Institute for Environmental Studies (NIES) installed the first super-computer in Japan, followed by another at the University of Tokyo in 1991, to conduct computer modeling. The University of Tokyo established the Center for Climate System Research in 1990; better computers had to be introduced to enhance research capabilities. Efforts were also made to publish research outputs from before 1990 so that they could be incorporated into the first IPCC assessment report (Council for Science, Technology and Innovation 2002).

Similarly, NIES started conducting other model development research concerning emission mitigation policies (Matsuoka et al. 1994). Development of the Asia-Pacific Integrated Model (AIM) began in 1990 to calculate the economic impacts of emission reduction policies. Scientific findings by Japanese scientists were incorporated mainly into the IPCC and other related international scientific activities but not directly into Japan's decision-making on climate change. Japanese people obtained information on climate change mainly from the mass media, and the media's coverage focused more on the IPCC's outputs than on those of Japanese scientists (Kawashima 1997).

Japan's role in INC negotiations and the adoption of UNFCCC

A formal multilateral negotiation to reach an agreement on a treaty started in early 1991, after the resolution "Protection of Global Climate for Present and Future Generations of Mankind" was adopted at the Plenary of the United Nations in December 1990 (United Nations 1990). The Intergovernmental Negotiation Committee (INC) was established by the resolution, and its first meeting was held in Washington, D.C., in February 1991.

During the following 18 months of intergovernmental negotiations on climate change, Japan basically favored a convention that would be agreeable to all major industrialized countries. Japan especially considered the United States to be the most important country to accept the agreement, because it alone was then responsible for nearly a quarter of global CO_2 emissions. In early negotiating meetings, the then ambassador Nobutoshi Akao and Mexican ambassador Edmundo de Alba-Alcaraz were elected as co-chairs of Working Group 1 of the INC, in charge of the commitment aspect of the convention.

At INC2 held in Geneva in June 1991, Japan proposed a "pledge and review" procedure (Akao 1993: 274–284). With this procedure, each country would pledge a set of national strategies and response measures to limit GHG emissions, as well as the emission goals that were expected to be achieved by using these strategies. A country's progress in fulfilling its pledge would be periodically evaluated by an international team of experts. The emission target itself would not be legally binding, so no penalty would be imposed if the target was not achieved.

Japan's rationale was that this was a good compromise between setting absolute targets, which some of the European countries wanted, and the position of the United States, which was reluctant to do so. This path would also be in Japan's interest because it was willing to set an emission target, but only at a lower rate than those set by many other developed countries. MITI and MOFA were proponents of this proposal, but not the EA, which wanted legally binding targets. Officials from MITI and MOFA explained internally that they had support from the United Kingdom and France for the proposal, but this was not entirely correct because these two countries held the position that all developed countries should stabilize CO_2 emissions in absolute terms by 2000 as a precondition. Japan's "pledge and review" proposal was criticized by many countries and environmental NGOs as being "hedge and retreat" – that is, a way to get around a strict emission reduction commitment (Bodansky 1993). MITI and MOFA decided that Japan would turn down the proposal at the INC3 meeting held in Nairobi in September 1991.

Japan was nevertheless supportive of the European position to include some kind of numerical emission targets in the convention. Having set its own national emission target in 1990, Japan had hoped that all industrialized countries would follow suit and set similar emission targets. It was also Japan's position to exclude GHGs other than CO_2, such as methane (CH_4) and nitrous oxide (N_2O) and not to incorporate sequestration of CO_2 by forests. At the time, data for these gases and sequestration activities were considered to be less certain than those for anthropogenic CO_2 emissions. In some ways, Japan appeared eager to confine the scope of the climate change problem to the area of energy use, rather than expanding the issue to forest conservation and land-use change.

Japan was also not enthusiastic about including issues related to development support for developing countries into the negotiations on the response to climate change. Both Japanese ministries and political leaders believed that Japan needed to help improve the global environment, but they were not sufficiently deeply concerned with the developmental aspects of this issue to consider that it should be included in the agreements at this point. They preferred that the climate change debate and the development debate be kept separate. This position obviously was not welcomed by negotiators from developing countries. For the developing countries group, dubbed the G-77 and China, insertion of principles such as "common but differentiated responsibilities" and the "right to development" were essential elements of the UNFCCC (Hyder 1994: 215).

During the 16 months of negotiations, most everyday Japanese were not well informed of the process that was ongoing at the international level. The negotiation meetings were addressed in the various media, but most people were not very aware of the consequences of the agreement or the relevance of the issue to their daily lives. The term "global warming" had some impact on people's perceptions of the global environment, but it was difficult to see the difference between global warming and some other environmental issues such as ozone depletion and acid rain. Japanese industries also appeared to be generally unaware of the negotiations. The text of the convention seemed to be dealing with general commitments that would have little impact on industrial activities.

INC4 was held in Geneva in December 1991; countries basically reiterated their respective positions, and made little progress. At this INC, however, the countries agreed to combine the two working groups' papers into one consolidated working document. INC5 was initially supposed to be the final meeting to agree on the text of UNFCCC. Toward the end of 1991, however, countries found it difficult to agree on a single text within the limited amount of time available. They agreed to extend INC5 and to convene again in May. The first round of INC5, held in New York in February 1992, saw progress in drafting the text. Most of the contentious issues were resolved, but conflicts remained about some sections of the draft. The most serious disagreement was on articles related to Annex I countries' (industrialized countries') commitments on emission mitigation. The United States remained unchanged in its resistance to targets and timetables on emission stabilization commitments.

An Extended Bureau Meeting was held in Paris in April 1992. Delegates from 20–30 countries participated in resolving the remaining issues. The wording for Article 4.2(a)(b) of the UNFCCC was a result of compromise language that was worked out by the United States and United Kingdom just before the second part of INC5 was to resume in May. The second part of INC5 resumed in New York, and the UNFCCC was finally adopted on 9 May 1992 (Borione and Ripert 1994). In the final stage of negotiations, Japan supported the final draft of Article 4.2(a)(b) because this was the most the United States was willing to support; Japan believed it was important for the United States to sign the Convention. Japan had hoped to play a role in creating the compromise, but it was the United Kingdom that finally drafted the compromise wording on emission targets.

The wording of the final draft was extremely complex. The Annex I Parties adopted the following wording:

> Each of these Parties shall adopt national policies and take corresponding measures on the mitigation of climate change, by limiting its anthropogenic emissions of greenhouse gases and protecting and enhancing its greenhouse gas sinks and reservoirs. These policies and measures will demonstrate that developed countries are taking the lead in modifying longer-term trends in

anthropogenic emissions consistent with the objective of the Convention, recognizing that the return by the end of the present decade to earlier levels of anthropogenic emissions of carbon dioxide and other greenhouse gases . . . would contribute to such modification, . . . (Article 4.2(a))

This wording allowed the interpretation that failing to achieve the stated stabilization objectives would not constitute a violation of the obligation to reach the emission reduction target. No agreement could be reached regarding emission levels beyond 2000, and no mention of them is made in the text of the Convention.

UNCED, held in Rio de Janeiro from 3 to 14 June 1992, was one of the most notable landmarks of the world's environmental movements. Some of the best-known multilateral environmental agreements, such as the UNFCCC and Convention on Biological Diversity, were opened for signatures. UNFCCC was signed by 156 countries, including Japan. The Japanese people were strongly influenced by the discussions of global environmental conservation at high-level sessions of UNCED.

Then prime minister Kiichi Miyazawa was absent from the meeting because of other domestic commitments. The Japanese Diet was discussing the Peace Keeping Operation Law enacted in June 1992, and Miyazawa prioritized the domestic political debate. He sent Environment Minister Shozaburo Nakamura to give a speech on his behalf (Kameyama 2002). In addition, Japan pledged to drastically expand its aid in the environmental sector to JPY900–1,000 billion (USD7.2–8 billion) during the five-year period starting 1992 (on a fiscal year [FY] basis). By the end of the fourth year (FY 1995), Japan had spent JPY980 billion, thus fulfilling its pledge one year ahead of schedule.

Japan's implementation stage of initial reaction to UNFCCC

After the adoption of UNFCCC in May 1992, Japanese policymakers turned to the domestic side to start the ratification process. Japan ratified the Convention in May 1993, becoming the twenty-first country to do so. Because the UNFCCC demanded that developed countries "adopt national policies and take corresponding measures on the mitigation of climate change" and that "returning to their 1990 levels" of GHG emissions was not considered to be an international commitment, it was not difficult for Japan to accept the commitments and ratify the convention. Most of the ministries took the emission reduction target as to be nonlegally binding in nature; this meant that no budget would be specifically allocated to achieve the target.

Rather than trying to reduce GHG emissions domestically, MOFA and the Ministry of the Environment (MOE, formerly EA), used the opportunity to expand their fields of activity into environmental diplomacy in the Asia-Pacific region (Kameyama 2002). MOE organized the "Asia-Pacific Seminar on Climate Change," beginning in 1991, at which experts from the Asia-Pacific region gather every year to discuss possible regional cooperation on climate change (EA 1996).

The Environmental Congress for Asia and the Pacific was also initiated by MOE, as an annual meeting at the ministerial level, to discuss a variety of environmental topics. Its first meeting was held in 1992.

At the domestic level, the *Basic Environment Law* was enacted in November 1993 (Government of Japan 1993). The primary purpose of this law was twofold: to integrate two fundamental laws related to the EA, the *Basic Law for Pollution Control* and the *Nature and Environment Conservation Law*; and to include new environmental issues that arose in the 1980s, most notably global environmental problems. The law covered not only climate change but also various other environmental policies, and it laid out four principles related to environmental policy in general.

In Articles 6 to 9, the law designated the responsibilities of the national and local governments, corporations, and citizens. The basic message was that all stakeholders in Japan are obliged to take some action toward an environmentally sound future. Article 15 requires the national government periodically to establish a basic environment plan. This was considered to be a way to be able to respond flexibly to the changing state of environmental issues. Article 22 discussed economic measures, but the description was rather fuzzy:

> The State shall appropriately conduct surveys and research on the effectiveness of implementing such measures with regard to prevention of interference with environmental conservation and on the effects of such measures on the Japanese economy; and should it be deemed necessary to implement such measures, the State shall make efforts to acquire the understanding and cooperation of the people with regard to utilization of such measures to prevent interference with environmental conservation. In this case, should such measures be implemented for global environmental conservation, the State shall consider international collaboration so as to appropriately ensure the effectiveness of such measures.

Articles 32–35 wrote about international cooperation. Specifically, Article 32 noted:

> The State shall make efforts to take necessary measures to secure international collaboration for global environmental conservation and to promote other international cooperation for global environmental conservation; and to assist conservation of the environment in developing regions and of those environmental features that are recognized to be of international value, which contribute to both the welfare of mankind and the healthy and cultured living of the Japanese people, and to promote other international cooperation for environmental conservation in developing regions.

Climate policies were reflected in the articles of the law referred to above, without elaboration on the concrete policies and measures required to reach the emission stabilization target set out in Article 4.2(b) of UNFCCC. This was, however, considered sufficient to ratify the UNFCCC, because the emission reduction target for 2000 was considered not to be legally binding.

Following the enactment of the Basic Environment Law, the first Basic Environment Plan was finalized in December 1994. It stipulated four key long-term goals: materials recycling, cohabitation with the ecosystem, participation by all stakeholders, and international cooperation. Climate change policy was addressed along with other local air pollutants in a section titled "conservation of the atmospheric environment." To respond to climate change, the plan defined the long-term goal to be in line with Article 2 of the UNFCCC: in the short term, the goal was to implement what had been written in the Action Plan to Arrest Global Warming of 1990. The Basic Environment Law, the Environment Plan, and the Global Warming Action Plan all lacked any real enforcement power to push the relevant ministries and agencies to implement more detailed policies and measures to actually reduce GHG emissions.

Decision-making factors – political leaders

A fundamental question raised in this chapter is, "Why did Japan suddenly turn *green* in the late 1980s to set an emission stabilization target for itself and engage in negotiations towards the adoption of UNFCCC?"

Since Japan had overcome a serious local pollution problem in the late 1970s, the environment was not a popular subject among Japanese political leaders until the late 1980s. Within the largest political party, the LDP, the environmental *zoku* was a small minority with limited political power. In Japan, a *zoku* is a group of politicians that are familiar with, and have expertise in, specific areas of interest and issues; they often have strong ties with related pressure groups and bureaucrats. There had been other *zoku*s, such as the construction *zoku* and the agriculture *zoku*; of these, the environmental *zoku* was one of the least powerful.

Even before the emergence of global environmental movements in the late 1980s, some politicians who were members of the Environment Working Group under the LDP had tried to push environmental legislation, such as laws pertaining to environmental assessment, but they had failed because of objections from other interest groups that favored expenditure on public works. Another case that was taken up by the Environment Working Group was draft legislation on controlling ozone-depleting substances in the late 1980s, because the draft designated MITI as the managing ministry. There was a battle between MITI and EA over distribution of authority between the two administrations, and, ultimately, the environmental *zoku* was able to pass the legislation by having both MITI and EA as ministries managing the law. Given these past problems, the members established the "Subcommittee on Global Environmental Problems" under the Environment Working Group in 1988 to begin to deal with a new global issue – global warming.

The source of the political power of the newly established subcommittee was Prime Minister Noboru Takeshita. Takeshita served as prime minister from November 1987 to June 1989. Takeshita, at the time of his inauguration, identified "international contribution" as one of the main pillars of his political agenda. Part of this effort including joining international "peace keeping operations," but the global environment was also included as part of Japan's contribution to international society.

As was discussed in previous sections, the Toronto G7 Summit in June 1988 was the first Summit meeting Takeshita attended, and it had a great impact on his perception of the global environmental problem. Although people's support for his administration was hurt by the introduction of a 3% consumption tax and the Recruit Cosmos scandal in the late 1988 and early 1989, Takeshita remained interested in the foreign-policy aspect of political issues. Takeshita established the "Inter-ministerial Council on Issues Related to the Global Environment" just before his retirement. Perhaps one of his motivations to be engaged in the global environment was to improve his public image, which had been damaged by the scandal that drove him into retirement (Murai 2001).

The next prime minister, Sousuke Uno, was in office only for two months. During his administration, he was invited to G8 Summit of the Arch in July 1989. Although some expected that climate change would be taken up at this meeting, the prioritized issues were related to China's Tiananmen Square protests, which had occurred 1 month before the summit. Japan was the only G8 Summit participant from the Asian region, and Uno was urged by other members of the group to work with China to improve its level of democracy. In August 1989, Uno experienced a personal scandal, and the LDP lost the nationwide election of the Upper House of the Diet. Uno resigned the same month.

Prime Minister Toshiki Kaifu occupied the office of prime minister from August 1989 to November 1991. He was not particularly known for either his "green" or his "brown" positions on the environment, but he was among the members of the "Takeshita group," and thus his political position was heavily influenced by former prime minister Takeshita. He became a member of the Global Legislators Organization for a Balanced Environment, established in 1989. Mayumi Morikawa, Setsu Shiga, Ishimatsu Kitagawa, and Kazuo Aichi – EA ministers appointed during the Kaifu administration – all contributed to the strengthening of political power to respond to global environmental issues. They managed to largely resolve tensions between EA and MITI over emission targets for 2000 by setting two-tiered targets.

The Gulf War started in January 1991, and more than 30 countries, led by the United States, formed an allied force to attack Iraq. The conflict between Iraq and Kuwait had started the previous year in July 1990, when Iraq invaded Kuwait, and the United States had requested that Japan contribute to the effort by dispatching military forces as well as by making a financial contribution. This was a controversial issue in Japan, because Japan had abandoned the use of military force outside its own territory, as stated in the Constitution drafted immediately after the end of World War II (Berger 1998). Prime Minister Kaifu made the decision not to send Japanese troops to the Middle East but made a financial contribution of US$13 billion to the Allied forces. Although the amount of financial support was not small, Japan's diplomatic position was criticized for being "checkbook diplomacy" because it had refused to send troops. Given this experience, LDP politicians thought that they needed to find other ways to contribute to international agendas. Implementing climate change measures was considered a solution to their problem.

Among many decisions formulated by government officials during the INC negotiations leading to the adoption of UNFCCC in May 1992 was the "pledge and review" proposal, which was tabled at the INC meeting in July 1991. The proposal was heavily criticized by the LDP because it was tabled at the INC meeting without prior consultation with either the EA or Minister of the Environment Kazuo Aichi, who had been leading the debate on climate change in Japan (Watanabe 2011: 42). For politicians, this incident was perceived partially as an example of the bureaucrats' monopoly over foreign policy. As a result, the LDP's Special Committee on Global Environmental Problems did not support MITI and MOFA's "pledge and review" proposal and ordered them to remove it.

Environment Ministers Ishimatsu Kitagawa and Kazuo Aichi were relatively active during the Kaifu Administration. For example, in July 1991, Aichi visited the United States and had a meeting with the then senator Al Gore. Aichi explained that Japan was in agreement with the European countries regarding setting CO_2 emission stabilization targets and asked the United States also to set absolute emission targets.

The next Prime Minister, Kiichi Miyazawa, served from November 1991 to August 1993. Personally, he was not all that enthusiastic about global environmental problems. Although 1992 was marked as the year of the global environment in Japan, and many politicians hoped to get more involved in the global environmental movement in one way or another, Miyazawa left most of the environmental agenda to his predecessors in Takeshita faction. In those days, the Keiseikai, a faction consisting of politicians supporting Takeshita, was the largest faction within the LDP. Meanwhile, Miyazawa was in a different faction (the Kochikai); this might have been one reason why Miyazawa avoided global environmental issues.

The "Council on Basic Environmental Issues" was set up in the LDP in February 1992 under the leadership of Takeshita. This was politically significant in Japan's decision-making, because "councils" were more powerful than "working groups" inside the LDP, and obtaining consensus within the LDP councils was critical in pushing certain agendas to be adopted by the Diet. Most of the members of the new council were from the Keiseikai faction. The council held three goals to aim at: Japan's contribution at UNCED; to establish the Basic Environment Law (which did not yet exist in Japan); and to introduce some sort of environmental tax.

During 1991 and 1992, major developed countries were engaged in discussions not only on climate change but also on sustainable development in general. They were particularly concerned about funding to promote sustainable development in developing countries. This issue was considered a key to the success in UNCED. The then UNEP Executive Director Maurice Strong asked Noboru Takeshita to host an international high-level conference in Tokyo in April 1992 to discuss this matter, and both Takeshita and Kaifu were happy to do so. The Tokyo Declaration on Financing Global Environment and Development was adopted at the conference, and it was perceived as a success by LDP members in laying the basis for funding institutions agreed upon at UNCED. In addition, the members of the high-level group involved in the preparation of this conference established a non-profit organization group called Global Environment Action (GEA).

At the same time, Miyazawa was busy with the Peace Keeping Operation (PKO) bill, which had been shelved for some time. Since the Gulf War began, the LDP had worked strenuously to pass the bill, which allowed the government to send the Self-Defense Force to aid the UN-PKO without making amendments to the Constitution. The modified bill passed through the Diet on the same day as when the UNCED was held. Although some of the LDP's political leaders thought the global environment was the best way to express Japan's "international contribution," others in the party considered PKO to be more important.

The influence of the GEA in Japanese decision-making could be seen after 1992. The *Basic Environment Law* was finally established in 1993. Since the 1960s, two primary pieces of legislation had been administered by the EA: the *Basic Law for Pollution Control* and the *Nature and Environment Conservation Law*. The former aimed at controlling pollutants from the industrial and transportation sectors, whereas the latter was geared toward nature conservation. These two laws reflected how the EA was constructed in the 1970s. The laws, however, were unable to deal with two emerging environmental problems, one related to consumer behavior and the other to global environmental problems. The aim of the Basic Environment Law was therefore to cover new types of environmental problems, while also integrating the two traditional areas of environmental administration. Article 5 of the law states that Japan should collaborate with other countries to protect the global environment. In Article 15, it also requires the government to regularly establish an Environment Basic Action Plan. The first Basic Action Plan was formalized in 1994.

When the new Basic Law was being designed, the introduction of a carbon tax or other environmental tax was also discussed. Prime Minister Miyazawa and the LDP supported inclusion of the tax, but it was strongly opposed by industry groups and MITI. A carbon tax was ultimately not included in the final version of the Basic Environment Law (Watanabe 2011: 42).

The latter half of the Miyazawa Administration was marred by another scandal. Shin Kanemaru, one of the most powerful LDP politicians of the time, was indicted of receiving JPY500 million in bribes from Sagawa-Kyubin, a fast-growing door-to-door delivery service company. Public opinion polls showed that support for the LDP was rapidly declining, and many LDP politicians left the party and established new political parties. The Diet election of July 1993 was an epoch-changing loss for the LDP, which had maintained its position as the largest political party since its establishment in 1955. Miyazawa resigned right after the election, in August 1993.

Morihiro Hosokawa of Nippon Shinto (New Japan Party) became prime minister in August 1993. His governance relied on the alliance of several small political parties, many of which consisted of former LDP members. His administration did not last long, and the next Prime Minister, Tsutomu Hata of Shinseito (the Newly Born Party), held office for only two months (April–June 1994). Neither of these prime ministers had an opportunity to make an appearance at a G7 or G8 meeting; this further encouraged them to focus on internal affairs rather than international agendas.

During this politically unstable period, government bureaucrats were primarily responsible for conducting administrative work, such as ratification of UNFCCC and implementation of related legislation. Because UNFCCC did not have legally binding emission stabilization targets and there was little or no political will from within Japan to work toward emission reductions, the government did not take additional action to curve the continuing growth of GHG emissions, even after it had ratified UNFCCC.

Decision-making factors – other subnational actors

Citizen groups

There was a general feeling of support for environmental issues among the Japanese people in the 1980s, and this helped Japan to take relatively environmentally sound positions on global environmental issues. The media published many articles related to the global environment, including ones about wildlife conservation and climate change, and people were generally enthusiastic about Japan's contributions in the international community. This mindset developed in a context in which the Japanese people were becoming economically wealthier, and this was especially true in the 1980s. The surge in the yen's value against the US dollar after the Plaza Agreement in 1985 served to increase the number of overseas Japanese tourists, and this also encouraged people to have a more global point of view. Many individuals looked forward to Japan's new role in saving the global environment.

The willingness of people to protect the global environment did not, however, stem from an understanding of basic science. Relatively few studies had been conducted in Japan on climate change before 1990. Most of the scientific findings were based on the IPCC reports, which the Japanese people viewed as very credible; they considered that Japan should be in line with other industrialized nations. At the time, the impact of climate change was expected to occur far in the future and mainly in developing countries. There was little discussion of the extent of the burden the Japanese people would need to bear if Japan were to achieve the stabilization emission target for 2000.

Most citizen groups in the 1980s had been formed earlier in the 1970s, during the battles against the government and industry about health hazards arising from local pollution. These groups had little interest in taking up global environmental issues as part of their activities. Some new NGOs established in the 1980s in Japan that dealt mainly with global issues were branch offices of international organizations, such as WWF Japan, established in 1971; Friends of the Earth Japan, established in 1980; and Greenpeace Japan, established in 1989. These "imported" NGOs benefited from obtaining information from their head offices in Western countries, but they had limited support in the local community. Among the very few domestic environmental NGOs was CASA (Citizens' Alliance for Saving the Atmosphere and the Earth), which originated as a local citizens' group that fought for local air pollution regulation in Osaka in the 1960s and 1970s. As early as 1994, CASA pointed out the failure of the "Global Warming Action Plan" of 1990 – that is, the action plan had no concrete commitments for the implementation of policies and measures to achieve the emission targets.

Industry

A similar situation to that of the citizens' groups was observed for Japanese industries and the business sector. The 1980s was the decade of the "bubble economy" in Japan, and with the yen surging against foreign currencies Japanese companies started to go abroad, buying natural resources for their own consumption and real estate. As Japanese companies came to be criticized for their activities in developing countries that harmed the local environment, they had become engaged in the process of considering environmental problems outside Japanese territory. Keidanren, an organization of Japanese industrial and business firms, announced its "Global Environment Charter" in April 1991. This charter was considered a fundamental document showing Japanese businesses' positions on the global environment. The charter included ten bullet points that outlined how Japanese companies should behave in their overseas activities (Box 2.2). There was no reference to global environmental problems, including climate change, in these bullet points. The Charter's "Guidelines for Corporate Action," however, included the category "Response to global problems." Companies were to "cooperate in scientific research on the causes and effects of such problems as global warming and they shall also cooperate in the economic analysis of possible countermeasures," "actively work to implement effective and rational measures to conserve energy and other resources even when such environmental problems have not been fully elucidated by science," and "play an active role when the private sector's help is sought to implement international environmental measures, including work to solve the problems of poverty and overpopulation in developing countries." As we look back, it is surprising to see the industry's willingness to take action even in the face of scientific uncertainty.

Box 2.2   Global Environment Charter (excerpt) (Keidanren 1991)

Environmental Guidelines for the Japanese Enterprises Operating Abroad

- Establish a constructive attitude toward environmental protection and try to raise complete awareness of the issues among those concerned.
- Make environmental protection a priority at overseas sites and, as a minimum requirement, abide by the environmental standards of the host country. Apply Japanese standards concerning the management of harmful substances.
- Conduct a full environmental assessment before starting overseas business operations. After the start of activities, try to collect data, and, if necessary, conduct an assessment.

- Confer fully with the parties concerned at the operational site and cooperate with them in the transfer and local application of environment-related Japanese technologies and know-how.
- Establish an environmental management system, including the appointment of staff responsible for environmental control. Also, try to improve qualifications for the necessary personnel.
- Provide the local community with information on environmental measures on a regular basis.
- Be sure that when environment-related issues arise, efforts are made to prevent them from developing into social and cultural frictions. Deal with them through scientific and rational discussions.
- Cooperate in the promotion of the host country's scientific and rational environmental measures.
- Actively publicize, both at home and abroad, the activities of overseas businesses that reflect our activities on the environmental consideration.
- Ensure that the home offices of the corporations operating overseas understand the importance of the measures for dealing with environmental issues, as they affect their overseas affiliates. The head office must try to establish a support system that can, for instance, send specialists abroad whenever the need arises.

The "bubble economy" popped in February 1991, and the economic activity of the Japanese business sector began to shrink. Even many business leaders did not realize the seriousness of the recession for several years, but gradually it appears that the industry point of view switched from a willingness to take action to the mindset that there was no room for them to respond positively to global environmental problems.

Summary

This chapter has summarized Japan's response to the climate change problem from the late 1980s through the first half of the 1990s. During this time period, political leaders viewed climate change as a perfect agenda domestically and diplomatically. Inside Japan, LDP politicians had been accused of various scandals and had lost credibility with the public. Tackling global environmental problems was a suitable way to prop up the politicians' tarnished images and regain political power in Japan. Internationally, Japan had been accused of being interested in economic issues only and not contributing much to other international issues. Global environmental problems – particularly those related to climate change – were seen as an agenda through which Japanese technology and financial power could make a substantial contribution to the international community.

This type of mindset was welcomed by MOFA, which was responsible for foreign policy. The position was also in line with that of the EA, which was attempting to find its new roles after overcoming the local pollution problems by the early 1980s. Even Japanese industries sent out some "green" signals to society.

Mass media also played an important role in diffusing information concerning events happening outside Japan. The public became more aware of global environmental issues and generally hoped that Japan would contribute to sustainable global development. Although environmental NGOs had yet to become influential, general public support of green politicians was clearly observed.

These observations illustrate that a large majority of Japanese stakeholders basically approved of taking positive actions on climate change, for many reasons. Few voices were heard to point out the scientific uncertainty associated with various aspects of climate change. Japanese people accepted climate change science as a fundamental basis of the debate, and the only issue was how much action was deemed to be necessary. The emission reduction targets did not cause political controversy, because the country's commitment under UNFCCC was not interpreted to be legally binding. The more Japan showed willingness to make international contributions, the more it was welcomed by various stakeholders in the country.

Note

1 Sato, Katsuaki (2011) *Matsuo Basho* [Basho Matsuo]. Tokyo: Hitsuji Shobo.

References

Akao, Nobutoshi (1993) *Chikyuuwa Uttaeru* [Agenda for Global Survival]. Tokyo: Sekai no Ugokisha (in Japanese).

Arrhenius, Svante (1896) "On the Influence of Carbonic Acid in the Air on the Temperature on the Ground," *Philosophical Magazine*, 251, 236–276.

Barney, Gerald O. ed. (1980) *The Global 2000 Report to the President*. Washington, DC: US Government Printing Office.

Berger, T. (1998) "From Sword to Chrysanthemum: Japan's Culture of Anti-Militarism," in M. Brown, S. Lynn-Jones, and S. Miller eds., *East Asian Security*. Cambridge: MIT Press, 300–331.

Bodansky, Daniel (1993). "The United Nations Framework Convention on Climate Change: A Commentary," *Yale Journal of International Law*, 18, 451–558.

Borione, Delphine and Jean Ripert (1994) "Exercising Common but Differentiated Responsibility," in Irvin M. Mintzer and J. Amber Leonard eds., *Negotiating Climate Change*. Cambridge: Cambridge University Press, 77–96.

Brenton, Tony (1994) *The Greening of Machiavelli*. London: Earthscan.

Broadbent, Jeffrey (1998) *Environmental Politics in Japan: Networks of Power and Protest*. Cambridge: Cambridge University Press.

Busby, Joshua W. (2010) *Moral Movements and Foreign Policy*. Cambridge: Cambridge University Press.

Committee on Japan's Experience in the Battle against Air Pollution (1997) *Japan's Experience in the Battle against Air Pollution*. Tokyo: The Pollution-Related Health Damage Compensation and Prevention Association.

Council for Science, Technology and Innovation (2002) *ChikyuOndankaKenkyu no Sai-zensen* [The Front Line of Global Warming Research]. Tokyo: Council for Science, Technology and Innovation.

Dauvergne, Peter (1997) *Shadows in the Forest: Japan and the Politics of Timber in Southeast Asia*. Cambridge: Cambridge University Press.

Environment Agency of Japan (EA) (1983) *Kankyô Hakushô* [Environment White Paper]. Tokyo: Government of Japan.

Environment Agency of Japan (EA) (1996) *Proceedings of the Sixth Asia-Pacific Seminar on Climate Change*. Tokyo: Government of Japan.

Ehrlich, P. (1968) *The Population Bomb*. New York: Ballantine Books.

Government of Japan (1990) *Action Plan to Arrest Global Warming*. Tokyo: Government of Japan.

Government of Japan (1993) *Basic Environment Law*. Tokyo: Government of Japan.

Group of Seven (1988) *Economic Declaration*, 19–21 June 1988. Available online at: http://www.g8.utoronto.ca/summit/1988toronto/index.html (accessed 15 September 2015).

Hardin, G. (1968) "The Tragedy of the Commons," *Science*, 162, 561–568.

Hyder, Tariq Osman (1994) "Looking Back to See Forward," in Irvin M. Mintzer and J. Amber Leonard eds., *Negotiating Climate Change: The Inside Story of the Rio Convention*. Cambridge: Cambridge University Press, 201–226.

Jäger, Jill and Tim O'Riordan (1996) "The History of Climate Change Science and Politics," in Tim O'Riordan and Jill Jäger eds., *Politics of Climate Change*. London: Routledge, 1–31.

Kameyama, Yasuko (2002) "Will Global Warming Affect Sino-Japan Relations?" in Hanns Günther Hilpert and Réne Haak eds., *Japan and China: Cooperation, Competition and Conflict*. New York: Palgrave, 140–157.

Kawashima, Yasuko (1997) "Comparative Analysis of Decision-Making Processes of the Developed Countries towards CO_2 Emissions Reduction Target," *International Environmental Affairs*, 2(2), 95–126.

Keidanren (1991) *Global Environment Charter*. Available online at: https://www.keidanren.or.jp/english/speech/spe001/s01001/s01b.html (accessed 31 July 2015).

Matsuno, Yu (2007) "Pollution Control Agreements in Japan: Conditions for their Success," *Environmental Economics and Policy Studies*, 8(2), 103–141.

Matsuoka, Yuzuru, Mikiko Kainuma, and Tsuneyuki Morita (1994) "Scenario Analysis of Global Warming Using the Asian-Pacific Integrated Model (AIM),"*Energy Policy*, 23(4/5), 357–371.

Miyamoto, Ken'ichi (2013) "Japanese Environmental Policy: Lessons from Experience and Remaining Problems," in Ian Jared Miller, Julia Adeney Thomas, and Brett L. Walker eds., *Japan at Nature's Edge: The Environmental Context of a Global Power*. Honolulu, HI: University of Hawai'i Press, 222–251.

Miyazawa, Kenji (1932) "Guskobudori no Denki" [A Biography of Guskobudori]. *Jido Bungaku*.

Meadows, D.H., D.L. Meadows, J. Randers, and W. Behrens III (1972) *Limits to Growth*. New York: Universe Books.

Murai, Kyo (2001) "Jiminto Shin Kankyozoku no Keisei to Hokai: Zoku Giin no Hensh [Formation and Destruction of Environmental Zoku in LDP: An Unique Species of zoku Diet Members]," *Kokusai Seiji Keizaigaku Kenkyu*, 7, 133–156.

Ohta, Hiroshi (2000) "Japanese Environmental Foreign Policy," in Takashi Inoguchi and Purnendra Jain eds., *Japanese Foreign Policy Today*. New York: Palgrave, 96–121.

Organisation for Economic Cooperation and Development (OECD) (1977) *Environmental Policies in Japan*. Paris: OECD Publications & Documents.

Prime Minister's Office (PMO) (1984) *Kankyo Mondaini Kansuru Yoron Chosa* [Public Opinion Survey on Environmental Problems]. Available online at: http://www8.cao. go.jp/survey/s59/S59–06–59–05.html (accessed 31 July 2015).

Research Institute of Innovative Technology for the Earth (RITE) (2003) *New Earth 21: Action Programme for the Twenty-First Century*. Available online at: http://www.rite. or.jp/English/lab/syslab/research/new-earth/about-new-earth-program/about-new-earth. html (accessed 15 September 2015).

Schreurs, Miranda (1996) *Domestic Institutions, International Agendas, and Global Environmental Protection in Japan and Germany*. Dissertation submitted in partial fulfillment of the requirements for the degree of Doctor of Philosophy at the University of Michigan.

Tabb, William K. (1995) *The Postwar Japanese System: Cultural Economy and Economic Transformation*. Oxford: Oxford University Press.

Takemoto, Kazuhiko (1991) *Japan's Initiatives on Global Warming*. Washington, DC: World Bank.

Tolba, M., K. Osama, A. El-Kholy, E. El-Hinnawi, M.H. Holdgate, D.F. McMichael, and R.E. Munn eds. (1992) *The World Environment 1972–1992: Two Decades of Challenge*. London: Chapman & Hall.

Ueta, Kazuhiro (2005) "Economic Implications of Pollution Control Policy in Japan," in Imura Hidefumi and Miranda A. Schreurs eds., *Environmental Policy in Japan*. Cheltenham: The World Bank and Edward Elgar, 86–101.

United Nations (1990) Resolution A/RES/45/212, *Protection of global climate for present and future generations of mankind*, the 71st plenary of United Nations on 21 December 1990.

Walker, Brett L. and William Cronon (2011) *Toxic Archipelago: A History of Industrial Diseases in Japan*, Wewyerhauser Environmental Books. Seattle: University of Washington Press.

Watanabe, Rie (2011) *Climate Policy Changes in Germany and Japan*. Abingdon: Routledge.

Wilkening, Ken (2004) *Acid Rain Science and Politics in Japan: A History of Knowledge and Action toward Sustainability*. Cambridge: MIT Press.

World Commission on Environment and Development (WCED) (1987) *Our Common Future*. Oxford: Oxford University Press.

3 COP3 and the Kyoto Protocol (1995–2002)

Overview

> *Meigetsu wo totte kurero to nakuko kana*
> (A beautiful full moon / give it to me / cries a child)
>
> Issa Kobayashi[1]

In the early nineteenth century, Kobayashi wrote of a small child importuning for the shining moon, like a giant diamond in the dark sky. In haiku, a beautiful full moon (*meigetsu*) is a term expressing the season of autumn, because Japanese people enjoy the clear moonlight in early autumn, when the heat and humidity of the hot summer have begun to retreat.

At the first Conference of the Parties (COP1) to the United Nations Framework Convention on Climate Change (UNFCCC) in 1995, Japan announced its willingness to host COP3 or one of the later COPs. An official decision was made at COP2 that COP3 would be held in Kyoto in late 1997. After two and a half years of intense negotiations from COP1, the Kyoto Protocol was adopted at COP3 in 1997. From one point of view, the adoption of the Kyoto Protocol with its legally binding emission reduction targets for each developed country could be seen as a remarkable success of the Japanese government as well as of the Japanese foreign policy in hosting COP3. From another perspective, however, the adoption of the Kyoto Protocol was the beginning of Japan's long-term suffering from the "6% emission reduction target between 2008 and 2012." Especially after the United States withdrew from the Kyoto Protocol in 2001, voices against ratification of the protocol were heard within Japan.

What was the purpose of hosting COP3, and how should it be evaluated? Was hosting a large multilateral conference in Japan and making it a great success as a part of Japan's international contribution a giant beautiful dream that, like the bright full moon in the haiku, could never be realized? Was Japan able to follow up on its success on the Kyoto Protocol after its adoption?

COP1 and the Berlin Mandate

Climate-change-related activities at both the international level and the domestic level in major countries were quite slow after the adoption of the UNFCCC in

June 1992. First, it took time for countries to ratify the convention; this was an unavoidable process in each country. In addition, the United States' political situation was not favorable for "green" initiatives in general. The United States had been one of the earliest countries to ratify the UNFCCC in 1992, but any further progress for more stringent climate policy became hopeless after the Republicans won a landslide victory in both houses of the US Congress in November 1994 (Oberthür and Ott 1999: 42–45).

Climate change negotiators at the time were aware of the success of other multilateral treaties such as the Vienna Convention on the Protection of the Ozone Layer and the Convention on Long-Range Transboudary Air Pollution, in which multilateral framework conventions worked as an institutional regime-building agreement and protocols were designed at later stages to determine specific numerical commitments for countries. The Alliance of Small Island States (AOSIS) and Germany had submitted draft proposals of a protocol for setting emission reduction targets for countries, expecting it to be adopted at COP1, but few countries, if any, were optimistic about agreeing to these proposals within less than a year of negotiations. Rather, more participants of COP1 shared the view that they were seeking a decision that would initiate another round of negotiations to reach an agreement on a new protocol under the UNFCCC. The protocol was assumed to include specific emission reduction targets for years beyond 2000.

COP1 took place in Berlin in March–April 1995. The core agenda was determining the adequacy of Article 4.2(a)(b) of UNFCCC. Two subparagraphs indicated the commitments of the Annex I countries (i.e., the developed country Parties), including nonlegally binding emission stabilization targets for 2000. There was no mention of years beyond 2000 or of whether or not the developed-country parties should continue to commit to actions without the participation of the developing countries.

During the COP1 meeting, the developed countries, including Japan, asserted that at least some emerging economies with a relatively large share of global greenhouse gas (GHG) emissions should start taking mitigation actions. Meanwhile, the developing countries strongly criticized the developed countries because GHG emissions in many developed countries, including Japan, were still increasing between the years 1992 and 1995. It was therefore difficult to expect those countries to be able to return their GHG emissions to the 1990 levels by 2000, as had been suggested by Article 4.2(b). Japan's GHG emissions had grown by 9% between 1990 and 1995, and the growth was expected to continue through 2000. The growth was also observed in per capita terms, so that Japan would not even be able to reach the per capita 2000 target set by the Action Plan to Arrest Global Warming in 1990. The developing countries considered that the developed countries were not taking the emission stabilization target under Article 4.2 seriously, and that stringent legally binding emission reduction targets should therefore be set for the developed countries for years beyond 2000 in the upcoming negotiated document. The developing countries also asserted that they had the right to catch up to the level of economic prosperity enjoyed by the Annex I countries; thus, any commitments to limit GHG emissions from these countries should be delayed until then.

After days of long discussion, the developing countries formed the "Green Group" and proposed a decision text that aimed at starting a negotiation to agree on a protocol that included emission reduction targets for developed countries only. Germany formed a consensus among other EU member countries that they would not require developing country commitments to be on the agenda of the negotiation. The JUSCANZ group (Australia, Canada, Japan, New Zealand, and the United States) opposed starting a negotiation on a protocol without the involvement of developing countries, but they did not wish to completely destroy the agreement.

The Berlin Mandate was adopted on the final day of COP1 in April 1995. It called for starting a process to enable the parties to take appropriate action for the period beyond 2000. This included strengthening of the commitments of the parties included in Article I to the convention in Article 4.2(a)(b), through the adoption of a protocol or another legal instrument. The legal instrument was to include an elaboration of policies and measures, as well as quantified limitation and reduction objectives (QELROs) within specific time-frames such as 2005, 2010, and 2020 for GHG emissions. It also stated that no new commitments would be introduced for non-Annex I countries (UNFCCC 1995).

Another decision was adopted on a pilot phase of Activities Implemented Jointly (AIJ). This scheme had been proposed by the United States and was supported by the developed countries as a means to collaborate among countries to reduce emissions jointly. The scheme was similar to joint implementation or the Clean Development Mechanism (CDM), both of which were later agreed upon under the Kyoto Protocol.

Japanese politicians and governmental officials were still eager to utilize the global environmental problem to enhance Japan's diplomatic influence as a mid-level power. Among the many types of global environmental problems, including biodiversity and ozone depletion, climate change debates was considered the best suited to sell Japan's energy-efficient technology and claim it as Japan's international contribution. At the ministerial segment of COP1, Sohei Miyashita, the then Environment Minister of Japan, stated that Japan was willing to host the third COP or one of the later COPs. Hosting the COP two years after COP1 meant that Japan would bear the responsibility as the host nation for the adoption of the protocol (Kawashima 2000). Environmental issues had also begun to appear on Japan's foreign policy agendas – in particular, in relation to overseas development aid to its Asian neighbors. This created even greater interest within the Japanese government to make constructive commitments in the negotiations to adopt a protocol in Japan. From another diplomatic point of view, no existing multilateral agreement had yet been named for a Japanese city, so the Japanese government expected that Kyoto, or some other city, would be included as part of the name of the resulting international legal document.

It is hard to identify any one key individual who led Japan's efforts to host COP3 (Ohki 2007). It seems that, by gathering unofficial information from a variety of sources, many bureaucrats and political leaders shared the vague idea that hosting an international conference on the global environment would be good for Japan, and few, if any, objection were heard to the idea. There were no objections when the Japanese cabinet ultimately approved hosting COP3 in May 1996, just before Japan made the official invitation for COP3 at COP2.

Ad Hoc Group on the Berlin Mandate (AGBM)

The negotiating process leading up to COP3 was called the "Ad Hoc Group on the Berlin Mandate" (AGBM). AGBM meetings were held eight times in 1995–1997. The most contentious debate in the negotiations was the same as it was in the 1991–1992 negotiations leading up to UNFCCC – how to set emission commitments for the Annex I countries, this time for years beyond 2000 – because the Berlin Mandate clearly stated that no new emission mitigation commitments would be set for non–Annex I countries.

The first three AGBMs were held with the goal not of negotiating legal texts but rather to share information and better understand the participants' views regarding the current status of energy use, the current level of energy efficiency, and policies and measures that were already in place. This type of dialog was an important part of the process to enhance credibility among countries and to recognize the respective countries' views of the issues at stake. During these meetings, Japan strongly emphasized that (1) it was already one of the most energy-efficient countries, and (2) it had relatively low per capita and per GDP emissions compared with those of other industrialized countries. Japan argued that its emission reduction target for the years beyond 2000 ought to be relatively less rigorous than those of other Annex I countries. It also insisted that setting the base year as 1990 would be unfair for countries such as Japan that had made great strides in improving energy efficiency in the late 1970s. The Japanese negotiators claimed the GHG emissions reduction in Germany and the United Kingdom since 1990 were due partly to the reunification of East and West Germany and a shift from coal to gas in the United Kingdom. They said that both of these circumstances had little to do with climate policy, and that these were "lucky" conditions that had put the EU in an advantageous position when 1990 was chosen as the base year.

AGBM4 was held in July 1996 in conjunction with the COP2 meeting held in Geneva. At these meetings, the EU suggested that a flat rate target for the QELROs was the most agreeable option because countries preferred different ways of "differentiation" – that is, setting different rates of emission reduction targets across countries according to certain criteria. Other developed countries, such as Japan, Norway, and Australia, asserted that some kind of "differentiation" was indispensable to reflect each country's national circumstances. On policies and measures, the EU considered it important for Annex I countries to coordinate policies to achieve emission reduction targets, whereas the United States wanted to set quantitative emission targets and let each country decide how to reach its own target.

The Second Assessment Report (SAR) of the Intergovernmental Panel on Climate Change (IPCC) had been published by then, and it more strongly indicated that there was a good chance that climate change was likely to occur because of the increase in the atmospheric concentration of anthropogenic GHGs. At this meeting, the United States, led by former Senator Timothy Wirth, then secretary for global affairs at the US State Department, accepted that the legal instrument to be agreed upon at COP3 would be a protocol. In addition, the United States also agreed to the idea of setting legally binding absolute emission targets for each country. The United States, at the time under the Clinton administration, proposed that full utilization of an emissions trading scheme would be applicable only if all parties, including developing countries, set absolute emission caps.

After the high-level portion of the meetings, ministers – particularly those from European countries and the United States – created the "Geneva Ministerial Declaration" (UNFCCC 1996a). It stated that "the parties recognize and endorse the SAR of IPCC as currently the most comprehensive and authoritative assessment of the science of climate change, its impacts and response options now available," with reference to essential elements of the report. It also went on to state that "significant no-regrets opportunities are available in most countries to reduce net GHG emissions" and, as a result, requested all ministers to "instruct their representatives to accelerate negotiations on the text of a legally-binding protocol or another legal instrument to be completed in due time for adoption at COP3. The outcome should fully encompass quantified legally binding objectives for emission limitations and significant overall reductions of GHGs within specified time-frames, such as 2005, 2010, and 2020."

This was the first time that the United States supported the wording "legally binding" in characterizing QELROs. Australia and some oil-producing countries strongly opposed giving full credibility to the IPCC reports. Thus, the COP could not adopt the declaration; it could only take note of it. Japan engaged in close discussions of the content of the declaration with the Australian government. Although Japan was hesitant to fully endorse the declaration, it did not oppose it either, primarily because it wanted to be viewed as a good host of the next COP. On the closing day of COP2, Japan reaffirmed its willingness to host COP3 in the ancient city of Kyoto.

After COP2, AGBM entered into the formal negotiating stage. During the negotiating process, Japan insisted on having differentiated targets for each party according to specified criteria, such as CO_2 emissions per capita or emissions per GDP. At AGBM5, held in December 1996, Japan proposed that a party could select from one of two different types of targets as defined in Box 3.1 (UNFCCC 1996b).

Box 3.1   Submission from Japan (UNFCCC 1996b)

Article 3

1   Each Party included in Annex I to the Convention shall select one of the following two quantified limitation and reduction objectives for its anthropogenic carbon dioxide emission by sources within the specified timeframes set out below:

(a) To maintain its anthropogenic emissions of carbon dioxide over the period from $[2000 + x]$ to $[2000 + x + [5]]$ at an average yearly level not more than p tonnes of carbon per capita, or,

(b) To reduce its anthropogenic emissions of carbon dioxide over the period from $[2000 + x]$ to $[2000 + x + [5]]$ at an average yearly level of not less than q percent below the level of the year 2000.

2   The Meeting of the Parties entrusts a study on anthropogenic emissions of greenhouse gases, other than carbon dioxide, not controlled by the Montreal Protocol, to the Subsidiary Body for Scientific and Technological Advice provided for in Article 9 of the Convention. Until such time as appropriate measures to limit and reduce emissions of such greenhouse gases are decided upon by the Meeting of the Parties on the basis of the study, each Party included in annex I to the Convention shall make as much effort as possible not to increase its emissions of such greenhouse gases.

(*) The concrete figures of x, p, q, should be developed.

As can be seen in Box 3.1, the proposal did not include particular values for x, p, or q, because Japan wanted to obtain supports from other countries on these formulae before getting into debates on concrete numbers. Many other Annex I countries supported the idea of differentiated emission targets, but desirable criteria for differentiation varied from one country to another. Australia insisted that countries that export a great deal of fossil fuel would be economically affected by emission reduction measures. It proposed that emission reduction targets should be differentiated by the degree of economic loss. Norway and Iceland proposed a formula that combined different criteria such as emission per capita and GDP per capita. It seemed to be difficult for these countries to agree on a single proposal to differentiate countries' emission reduction targets.

In March 1997 during AGBM6, the EU member countries agreed on their common negotiating position: "The Annex I Parties shall, by the year 2010, reduce their emissions of three types of GHG (CO_2, CH_4, and N_2O) by 15% from 1990 levels". The EU insisted that all Annex I Parties should commit to the same reduction rate of 15%, saying that an agreement on criteria for differentiation would be hard to achieve. Among the EU countries, however, the rates of emission targets were diverse – from a 40% increase for Portugal to a 30% reduction by Luxembourg. Japan, along with other non-EU countries (Norway, Australia, and New Zealand), argued that the EU proposal was inconsistent on the matter of differentiation because there was differentiation internally within the EU countries. They also noted that the reduction rate of 15% was unrealistic for many non-EU developed countries and that even the EU member countries were able to agree on an emission reduction of only up to 10% and that no explanation was given as to where the additional 5% reduction was going to be allocated. Meanwhile, these countries for differentiated targets were unable to present their own negotiation positions on concrete figures for emission reduction targets, and Japan was under pressure to propose specific targets for future emissions.

After AGBM6, the relevant ministries in Japan began to formulate ideas for a proposal, but it was difficult even to agree on a business-as-usual scenario for emissions beyond 2000. During AGBM7, which was held in July–August 1997, representatives from the Ministry of International Trade and Industry (MITI) confidentially suggested that p and q from the previous Japanese proposal be set at 3 (tons of carbon emitted per capita) and 0 (percent reduction from the 2000 level), respectively, for the United States and EU governments. The United States was opposed to any proposal that included a per capita basis, claiming that it would send a message to developing countries that all countries would be allowed to emit up to that level. The EU was also opposed, but for another reason – the proposal would only stabilize Annex I countries' total emissions, and this would not be enough to meet what the EU expected for the new protocol, to reduce the total GHG emissions from Annex I countries to a certain extent (Tanabe 1999).

Industry groups in Japan were also beginning to pay more attention to the negotiating process. They were aware that industry groups in other countries were also keeping a close watch on the process and taking actions before government actions so as not to be negatively affected by the final agreement reached at COP3. Keidanren (the largest Japanese industry organization) started internal discussions on how to respond to the climate change debate. Industries generally wanted to avoid any firm regulations of their activities. Before being regulated by the government, the industry group decided it would be better to take action. The Keidanren Voluntary Action Plan was published in June 1997. Altogether, 137 organizations from 36 sectors set emission reduction targets, many of which were not absolute reduction targets, but rather relative targets, such as emissions per unit of production (Keidanren 1997). In a way, the Keidanren was sending a message to the government, pressuring it not to introduce any additional measures after COP3.

AGBM7 was held in Bonn in July 1997, with less than six months remaining until COP3. The meeting divided the agenda into four groups: QELROs; policies and measures; Article 4.1 of UNFCCC (commitments for all Parties); and institutional issues. Countries reiterated their positions at the meeting, and little progress was made.

Outside Japan, the United States also found in itself in the midst of an intense conflict between those who wanted COP3 to make progress in tackling the climate change issue and those who were opposed to taking any action to curb GHG emissions. In July 1997, Senate Resolution 98 (the Byrd–Hagel Resolution) was passed in the Senate by a vote of 95–0. Although the resolution was nonbinding in nature, it stipulated that the United States refrain from signing any agreement at COP3 that did not also impose commitments on developing countries as well as industrialized countries and that any proposal could potentially have a serious impact on the US economy (Harris 2000). This resolution became a heavy burden on the Clinton-Gore administration in terms of climate change negotiations.

Japan's proposal for QELROs

Japan's two leading ministries, the Environment Agency (EA) and MITI, also started preparations for an internal debate on Japan's emission reduction target for 2010. The Joint Conference of Relevant Advisory Councils on Domestic Measures Addressing the Global Warming Issue was set up in July 1997. MITI's Industrial Structure Council worked between April 1996 and March 1997 to investigate emission mitigation potentials by improving energy efficiency, and it was in a position that no further emission reduction was unrealistic in Japan. The Central Environmental Council under the EA had been reviewing the Basic Environment Plan and had commented that additional measures would be necessary to meet the emission stabilization commitment for 2000. It also called for revision of the Action Plan to specify policy instruments to further reduce GHG emissions in Japan. Members of the two councils gathered at the Joint Conference, but the groups insisted on their respective viewpoints, and no agreement was reached.

In addition to the debates in the Joint Conference, the EA, MITI, and Ministry of Foreign Affairs (MOFA) were internally discussing the same issues. As COP3 approached, Japan was feeling the pressure to submit a concrete proposal for an emissions reduction and limitation target, but because of divergent interests at the domestic level the Japanese government remained unable to present a national position. In the run-up to COP3 in August and September of 1997, the relevant ministries continued to hold meetings to draft an acceptable government position (Takeuchi 1998; Tanabe 1999).

Late in September 1997, high-level officials of the government met to finalize discussion on the Japanese position on emission targets. MITI argued that Japan's target could be an emission limitation target rather than a reduction target, and that stabilization of emissions (reducing emissions to 1990 levels by 2010) would be the most that Japan could hope to achieve because of strong opposition from industry. For MITI, which was responsible for domestic energy supply and industrial policy, it was obvious that Japan, with one of the lowest per capita emission rates among developed countries, would find the task of reducing GHG emissions to 1990 levels more arduous than others. MITI thus sought some kind of differentiated targets. At the same time, it was crucial for MITI to gain the United States' participation from the point of view of Japanese industry remaining internationally competitive. Japanese officials thought the United States would not be able to agree on any stringent emission reduction target, even if Japan wanted to take the lead in the negotiation to achieve any ambitious outcome for COP3.

Meanwhile, the EA expected that hosting COP3 would be an excellent opportunity for Japan to determine an ambitious emission reduction target. In looking at the climate change debate from a global environmental perspective, the EA argued that Japan should propose a draft protocol that incorporated emission reduction targets that were ambitious enough to avoid serious adverse climate change impacts but, at the same time, were realistic enough to be agreeable. The EA used an economic model called the Asia-Pacific Integrated model (AIM) to claim that a target rate of a 6%–8% reduction from 1990 levels by 2010 could be reached if sufficient additional measures were implemented. AIM was developed in 1990 by the National Institute of Environmental Studies (NIES) to respond to questions that arose from government officials regarding the cost of emission mitigation, impact of emission mitigation policies on the Japanese economy, and any likely adverse impacts of climate change. This model was the only comprehensive energy-economic model in Japan at the time of negotiations leading up to COP3. MITI, with ample support from industry, criticized the outputs of AIM by pointing out that the underlying assumptions used were different from those used by the industry groups.

MOFA's main objective in hosting COP3 was to conclude a successful multilateral meeting in Japan, which, in this case, meant the adoption of the protocol without mishap. To do so, Japan needed to persuade the United States to agree to adopt the protocol. Inevitably, Japan would have to coordinate with other countries to bend to the wishes of the United States. Conversely, if Japan wanted to enhance its status as a leader in Asia, it needed to set a good example on global environment issues. Given that the EU had been calling for a 15% flat rate reduction target since March of the same year, MOFA felt that a reduction of 5% at the very least was necessary and argued for a target of 6.5%. After communicating with other major countries through unofficial channels, MOFA came to the opinion that a reduction of at least 5%–6.5% was needed to call COP3 a success. Furthermore, from Japan's point of view, if Japan were to agree to a 5% reduction, then Annex I countries as a group would need to agree on an even deeper reduction, because Japan continued to insist on the importance of differentiation.

After intensive consultations with various interested parties, MITI conceded to agree to stabilization at the 1990 level. This was, however, still far from the EA's 6%–8% reduction proposal and MOFA's 6.5% reduction proposal. Politicians were not directly involved in the discussions. Rather, the cabinet secretariat played a role as a mediator among the three ministries. He agreed that at least a 5% reduction was necessary for Annex I countries as a whole, but some kind of differentiation was necessary to reflect each country's national circumstances (Takeuchi 1998). They finally agreed on a proposal, which was a basic reduction of 5%, with a proviso for exceptions for certain countries, including Japan. This proposal was actually a reflection of "the differences that existed among the major domestic players and interpretations of what might be acceptable internationally" (Schreurs 2002).

In October, Japan officially submitted its proposal on an emission target as shown in Box 3.2. It proposed a 5% base reduction rate for each country. Countries with the certain conditions were allowed to apply to any one of the alternative reduction rates as noted in Box 3.2.

Box 3.2   Japan's Submission on QELROs (excerpted from UNFCCC 1997)

(a) For a country of which emissions per GDP in 1990 (A) are less than the emission per GDP of all Annex I countries in 1990 (B): Alternative Reduction Rate (%) = 5% × (A/B);

(b) For a country of which emissions per capita in 1990 (C) are less than the emission per capita of all Annex I countries in 1990 (D): Alternative Reduction Rate (%) = 5% × (C/D);

(c) For a country of which population growth from 1990 to 1995 exceeds the population growth of all Annex I countries for the same period, the higher growth of population should be considered in deciding the target of the country. Concrete formulation of this alternative reduction rate is to be developed.

Under no circumstances shall any country's emissions exceed its 1990 levels.

The first and second options reflected Japan's concerns about its own disadvantage owing to its high level of energy efficiency. The last option reflected a concern of the United States and was included so that the proposal would be more agreeable to the US government. Using this formula, Japan's own emission reduction target would have been 2.5%.

Even though Japan had attempted to consider the United States' national circumstances and incorporate its concerns into its proposal, the United States did not support differentiated targets and instead preferred a simple flat-rate emission reduction target for all Annex I countries. The final AGBM before COP3, AGBM8, was held in October 1997. In early October just before the AGBM meeting, US president Clinton had invited industry representatives, scientists, and environmental groups to a White House conference on climate change. Drawing upon the discussions at this meeting, the United States finalized a proposal for QELROs, which it announced on the opening day of the AGBM8 meeting. The United States proposed stabilizing net GHGs emission at 1990 levels sometime between 2008 and 2012. This proposal stipulated certain preconditions, including concessions on an emission trading system, whereby countries could use emissions trading and joint implementation schemes. Another condition sought by the United States was some type of meaningful participation by the developing countries, which had been excluded from any emission reduction obligations by the Berlin Mandate.

The agenda was again categorized into the four groups used at AGBM7. Developed countries were still not able to agree on the matter of differentiation QELROs. Now that the United States had proposed a flat stabilization target for all countries, country groups that had fought for differentiation thought they were being put at a disadvantage. In addition, these countries were also not being able to agree among themselves how differentiation should actually be reflected in the respective emission reduction targets (Torvanger and Godal 1999). Once again, these countries criticized the EU for supporting flat-rate targets while still using differentiation within the countries of the EU.

The types of gases being regulated also became an issue. Japan and the EU supported regulation of three gases: CO_2, methane (CH_4), and nitrous oxide (N_2O). They selected these gases primarily because of the availability of relevant data. Other countries, including the United States, preferred six gases by also including hydrofluorocarbons (HFCs), perfluorocarbons (PFCs), and sulfur hexafluoride (SF_6), as well as sequestration by sinks, which included land use, land-use change, and forestry. These underlying conditions significantly affected the level of QELROs, so these issues became increasingly important as countries started comparing emission targets.

Progress was made in the area of the base year and target year. Countries agreed to use 1990 as the base year except for countries classified as "economy-in-transition," and they also agreed to use five consecutive years as the commitment period, rather than setting a single year such as 2010 as the target year. Other progress was observed in the area of flexibility mechanisms. The EU initially had opposed the use of emissions trading for two reasons: (1) it felt that most efforts for emission mitigation should be done domestically, and (2) it thought that it was to too early in the process to establish global trading mechanisms. During the AGBM8 meeting, however, the position of the EU on emissions trading became more flexible, mainly because of strong pressure from the United States on this point. Meanwhile, the developing countries were strongly opposed to the use of emissions trading because they thought this type of mechanism would be used as a loophole for developed countries to buy emission permits from developing countries. The Berlin Mandate had eliminated developing countries from setting emission caps under the initial commitment under the Kyoto Protocol, but, even at that time, most countries believed that at least some non–Annex I countries would have to be included in Annex I in the near future.

There was little reaction from the Japanese government or industry groups on issues related to emissions trading. The use of flexibility mechanisms would be good for countries such as Japan because it could make achieving specific emission reduction targets less costly, and this could lead participants to agree to a more ambitious emission reduction target than they otherwise would. On the other hand, such instruments could be utilized to avoid any actual emission reduction in Japan because it could achieve its emission reduction target by purchasing emission allowances from abroad. Generally speaking, emissions trading was a relatively new concept in Japan, and not many Japanese expected this type of carbon market to be established in the near future, particularly from a technical standpoint, including the establishment of a system to register and trade emissions.

Kyoto Conference (COP3)

The 3rd Conference of the Parties was held from 1 to 10 December 1997 at the Kyoto International Conference Center, located in the northern part the city. Hiroshi Ohki, Japan's Environment Minister, was assigned the role of president of COP3, and Keizo Obuchi, the then MOFA Minister, gave a speech welcoming the participants and emphasizing the importance of tackling the climate change issue.

All of the major parties came to COP3 with their respective proposals. During the actual negotiations, however, the details of the proposals needed to be hammered out before the figures for emission levels could be negotiated, because different preconditions were used for the various targets. These preconditions included the type of GHGs covered (whether or not to include trace gases such as HFCs, PFCs, and SF_6 in addition to CO_2, CH_4, and N_2O); whether or not, and how, to include sequestration by sinks; whether or not to allow the use of flexibility mechanisms such as emissions trading and joint implementation; and whether or not to include the developing countries. Japan's proposal on differentiation of emission reduction targets became relatively less important during these negotiations, as many these other components started to come into play.

Japan had proposed consideration of three gases only, so it had to reconsider how the emission target might change if this underlying condition changed. Emissions of the three additional gases had been increasing in Japan since 1990. Among the three, increases in HFC emissions were the most serious. HFC emissions were increasing as a response by industries to replace ozone-depleting substances, as was agreed upon under the Montreal Protocol on Substances that Deplete the Ozone Layer. The protocol prohibited use of chlorofluorocarbons (CFCs) and recommended a shift to HFCs, a non-ozone depleting substance but a GHG with a high global warming potential. Japan considered that the base years for the three additional gases needed to be changed if HFCs were to be included in the Kyoto Protocol.

In addition, Japan and the EU had proposed that only emissions and not sequestration by sinks be targeted in the agreement. This idea had come to be called the "gross approach" by the time of COP3. Issues concerning sequestration by sinks had not received much attention during the AGBM meetings, because the negotiators were too busy negotiating differentiation or flat-rate targets. Many uncertainties also remained in this area at that time. At the least, the degree of precision of data on CO_2 emissions from fossil fuel combustion was considered to be far too different from that of sequestration by sinks, and this would make it difficult for governments to add or subtract these figures on the same basis. New Zealand, Australia, and Canada, supported by the United States, emphasized the significant role of forests in climate change, and they called for targets that accounted for sequestration by forests and land-use changes. This approach came to be known as the "net approach." The net approach was technically difficult, because there were no data available for sequestration by sinks in 1990. Nonetheless, the Annex I countries discussed practical ways to include this approach into QUELOs. The discussions revolved around issues such as the types of forest-related activities to be taken into account. Some countries proposed that the scope should be limited to afforestation and reforestation after 1990, but other countries argued that forest management and conservation should also be accounted for.

The negotiations also included discussions of three different ways to transfer emission allowances. Initially, the United States wanted to have emission caps for all countries, including developing countries, but this was strongly opposed by the group of developing countries. Meanwhile, Brazil proposed the "clean development fund," a funding mechanism scheme in which Annex I countries that did not meet their emission reduction target would face financial penalties. The money paid by non-compliant countries would be deposited in this new fund, which would be used to fund GHG reduction activities in developing countries. This idea was not popular with Annex I countries, but the United States and Brazil worked together to jointly propose the Clean Development Mechanism or CDM, which was similar to the clean development fund, although the money paid would not be a penalty but a payment for the purchase of emission credits (Grubb et al. 1999; Oberthür and Ott 1999).

Discussions on the US proposal on the use of emissions trading, joint implementation, and credit-based mechanisms faced many obstacles, even after negotiators agreed to exclude developing countries from emissions trading and joint implementation. The EU still held the position that most emission reduction efforts should be made domestically. In addition, developing countries retained their opposition to any types of carbon market, even if these schemes were used only among Annex I countries. They were concerned about the establishment of a regime in which rich countries could continue emitting GHGs by purchasing emission allowances from poorer countries.

Japan was situated somewhere in the middle of the conflict arising from discussions on the use of flexibility mechanisms. It would be helpful for Japan to be able to buy emission allowances if it were unable to meet its emission reduction target. It was also important for Japan to do what was necessary to keep the United States in the negotiations. On the other hand, the Japanese government had already agreed internally about the distribution among ministries of responsibility for emission reduction rates, and the use of flexibility mechanisms had not been considered by all the relevant actors. If Japan were to accept a greater emission reduction target under the condition that it could also use these flexibility mechanisms, then the negotiators in charge of flexibility mechanisms would have to consult with the Ministry of Finance (MOF) in advance to secure a budget sufficient to acquire the allowances; this was considered to be an additional hurdle internally. As the host country, Japan also did not want to unnecessarily aggravate its relationship with the developing countries' group. Japan therefore remained relatively quiet on this issue.

An article concerning the protocol's entry into force was another independent debate, and it was of great concern to Japan. Japan held the strong view that the United States, which was responsible for more than one-third of GHG emissions from Annex I countries, had to be a party to the protocol. Therefore, with the support of the EU, Japan asserted that the entry into force should be linked to US participation. Several concrete figures for percentages of emissions among Annex I countries were proposed to formalize the EU's and Japan's idea as a condition for the protocol to enter into force. The United States maintained the position that ratification of the protocol by a specified number of countries would suffice.

Japan prioritized the assurance of US participation in the protocol to the point that it began to support the US position, rather than its own. Japan proposed an emission stabilization target because it considered that the United States would never accept an emission "reduction" target. Japan also proposed voluntary actions for the developing countries to take into account the conditions of the Byrd–Hagel resolution, which was adopted by the US Senate in July 1997. Not surprisingly, the group of developing countries (G77+China) protested strongly, and Japan found itself caught between the two camps. The United States also asserted full use of emissions trading and crediting mechanisms, which were also opposed by the G77+China group. They argued that such mechanisms prevented developed countries from seriously reducing emissions at home. In the end, the G77+China and the United States made a deal among themselves: the United States accepted the idea of no emission commitments for developing countries but succeeded in getting international emissions trading included in the text.

During COP3, all relevant ministries and agencies within the Japanese government continued to review the various proposals. They estimated that a mere 1% reduction target would be equal to the installation cost of two new nuclear power plants, or about JPY900 billion in monetary terms. Toshiaki Tanabe, ambassador in charge of the global environment in MOFA, was the head of the Japanese delegation. As the time drew near the final day of COP3, he had to consult with Prime Minister Hashimoto to determine an emission reduction target for Japan. The high-level ministerial part of the meetings began on 8 December. US vice president Al Gore gave a speech, stating that the US delegation would show flexibility in the negotiations, suddenly shifting from its firmly stated position of 0% stabilization to a 7% emission reduction target. This came as a great surprise for the Japanese negotiators, because most of the bureaucrats in MITI strongly believed that the United States would never change its position from 0%. Prime Minister Hashimoto made phone calls to President Clinton, President Chirac of France, and other heads of state to explore the scope of a potentially agreeable solution.

A critical moment in the negotiations occurred during a closed meeting called for by AGBM Chair Ambassador Estrada on the afternoon of 10 December. Delegates from Australia, Canada, the EU, Japan, New Zealand, Poland, Russia, and the United States (from the Annex I group) and China, India, Samoa, and Tanzania (from the non-Annex I group) were in attendance. Up to this point, Japan had remained firm on its proposed 5% reduction target, but, in that meeting, Japan said that it would accept a 6% reduction target. At this point, Japan, the United States, and the EU respectively agreed upon fixed target reductions of 6%, 7%, and 8% of the baseline 1990 levels. Prime Minister Hashimoto was personally in favor of an environmentally sound position, and he reportedly instructed the Japanese delegation in Kyoto to accept the 6% target (Kobayashi 2012). However, this 6% target was acceptable only with one important precondition: that Japan would be able to count emission sequestration by land use, land-use changes, and forestry (LULUCF) as 3.7% of its emission reduction during the first commitment period of the Kyoto Protocol. At the time of the informal consultation, neither delegates from the other countries at the meeting nor Ambassador Estrada made objections to the proposal (Ohki 2007).

In the final hour of the open whole-group AGBM meeting, Ambassador Estrada heard objections from some delegates against the inclusion of LULUCF in the first commitment period and suggesting deletion of the sentence including LULUCF. Japan objected to this suggestion, but Ambassador Estrada used his gavel to formally agree on the deletion. With a sense of great urgency, members of the Japanese delegation informally consulted with other delegates and also asked Ambassador Estrada to consider reopening the discussion on the articles related to LULUCF. These consultations went into the early hours of 11 December. Japan asserted that it would have to reject the 6% reduction target if the use of LULUCF was not accepted during the first commitment period. Finally, the parties came to an agreement and a sentence was added to Article 3.4, stating, "A Party may choose to apply such a decision on these additional human-induced activities for its first commitment period, provided that these activities have taken place since 1990." This revision was later accepted in the final meeting of the whole group of delegates.

Negotiations were finalized in the morning of 11 December 1997, and the Kyoto Protocol was adopted. Article 3 defined the commitments on emissions limitations and reductions for the Annex I Parties. The first commitment period was set as a timeframe of five years between 2008 and 2012. The countries' QEL-ROs were listed in Annex B of the document, in which the EU, the United States, and Japan committed to reduce their emissions by 8%, 7%, and 6%, respectively, from the 1990 levels. Parties could use 1995 as the base year for HFCs, PFCs, and SF_6. Countries were allowed to consider sinks under Articles 3.3 and 3.4. Implementation and coordination of policies and measures were reflected in Article 2. Joint fulfillment of commitments (a rule proposed by the EU) was outlined in Article 4. Three ways of acquiring emission permits from other countries – joint implementation, CDM, and emissions trading – were allowed in Articles 6, 12, and 17, respectively.

In many ways, Japan got most of what it wanted: participation of the United States, differentiation of emission reduction targets among Annex I countries, and inclusion of sequestration. Its major trade-off was to accept the 6% reduction target.

COP4 and the Buenos Aires Plan of Action

Japan committed to a 6% emission reduction target in the Kyoto Protocol. GHG emissions in Japan, however, had been on the rise since 1990. In 1990, the level was 1.27 billion tons CO_2 equivalent, but by 1998 it had increased to 1.35 billion tons – a 6.0% increase from the 1990 level. It was quite clear that additional policies and measures would be necessary to achieve the Kyoto target.

Soon after COP3, the Japanese government instituted new procedures that required the implementation of the Kyoto Protocol. As the host of COP3, Japan felt a responsibility to enter the protocol into force as soon as possible, and therefore it took action immediately after the adoption of the protocol. The Global Warming Prevention Headquarters was established in the cabinet only ten days after COP3 under Prime Minister Hashimoto's initiative, so that specific and effective measures would be promoted for the implementation of the Kyoto Protocol (Government of Japan 2002).

An important issue facing the Global Warming Prevention Headquarters was how to distribute the responsibility to achieve the 6% reduction target among the various sectors – that is, the different ministries and government agencies. The allocation of responsibility had been discussed to some extent during negotiations at COP3, so matters related to the distribution of responsibility basically required only minor adjustments. The core issue was whether or not a new separate guideline focusing specifically on climate change was needed and, if so, whether or not the distributed responsibility should be framed as legally binding obligations. The EA, with the support of politicians who supported environmental issues (the environment-*zoku* or clan), emphasized the need to establish an independent guideline. MITI, with the support of industry groups and business-*zoku* politicians, argued that revisions to the *Law Concerning the Rational Use of Energy* and other existing laws were sufficient to respond to the commitments made under the Kyoto Protocol. The final agreement represented a compromise: a new guideline was approved, but there was no specific requirement that the energy and industrial sectors must control their GHG emissions. The industry groups were also successful in preventing the expansion of the EA's authority over industry activities (Watanabe 2011: 50).

The headquarters drew up the *Guideline of Measures to Prevent Global Warming* in June 1998. It was characterized as a set of rigid rules that clearly allocated responsibilities to various sectors to reach the 6% emission reduction commitment as a whole, but the guideline lacked any significant enforcement power to drive relevant Japanese stakeholders in the direction of emissions reduction. Emissions in Japan were expected to increase by more than by 20% by the time of Kyoto Protocol's first commitment period if no climate change mitigation policies were put in place. Reductions in this "without-policy" situation were to be achieved through further improvements in the use of energy and a shift to nonfossil fuels. A 2.5% reduction was to be achieved in areas related to energy combustion. This included emission reduction in industrial sector through improvements in energy efficiency and by using less carbon-intensive energy, as well as emission reduction in residential, commercial, and transportation sectors by the introduction of innovative technologies and changes in social behavior. The figure also took into account 0.5% reduction to be achieved by reducing emissions of GHG gases other than CO_2, such as CH_4 and N_2O. Emissions of HFCs, PFCs, and SF_6 were allowed to increase by an approximate 2% of total GHG emission level in the base year (1990, and 1995 for HFCs, PFCs, and SF_6), because these gases were expected to grow rapidly. Other reductions were expected to be achieved by sequestration by LULUCF-related activities (3.7%) and by acquiring emission permits through mechanisms such as international emissions trading, joint implementation, and the CDM (1.8%) (Government of Japan 1998).

To facilitate implementation of the guideline, the *Law Concerning the Promotion of the Measures to Cope With Global Warming* (Global Warming Measures Promotion Law) was established in October 1998. This law did not stipulate emissions reduction as its major obligation, because the Kyoto Protocol had not yet entered into force. Rather, this law reconfirmed the responsibilities of all relevant stakeholders – that is, the central government, local governments, businesses, and citizens. In addition, the law called for formulation of Action Plans by the central government, local governments, and business enterprises.

One of the important roles of the government under the law was to estimate national emissions and removal by sinks to implement its reporting and inventory requirements under the UNFCCC. The data had to be made available to the public on an annual basis. Another central obligation imposed by the law onto private companies was the obligation to report their GHG emissions annually, although this law imposed the obligation only on large emitters, which had to report energy-related CO_2 emissions and other GHG emissions annually. Any person was then able to request the disclosure of the submitted data. If companies failed to report or reported incorrectly, they were subject to financial penalties (Takamura 2012).

After this law was enacted, the Central Environment Council adopted the *Basic Guideline for Mitigation of Climate Change* in March 1999 to address specific policies and measures that were not dealt with in the law. The guideline said that the use of nuclear power was to be enhanced "while at the same time getting recognition of the public," because there had always been strong arguments against the use of nuclear power in Japan. The *Law Concerning the Rational Use of Energy*, initially enacted in 1979, was amended in June 1998 and entered into force in April 1999. It further strengthened the regulation of the energy efficiency of electric appliances. The revision introduced the "Top Runner Approach," a type of regulatory measure that set energy efficiency standards at a level slightly above the average of each product category. The standard was to be achieved by a target year, and the standards were planned to be gradually shifted upward to the level of the best-performing product. Industries were required to report their emissions to the government on an annual basis. Other policies, such as the introduction of domestic emissions trading or a carbon tax, were also investigated, but no official actions were taken.

The *Basic Guideline for Mitigation of Climate Change* seemed to be a good first step in taking action to meet the Kyoto emission target, but it was based on two crucial assumptions that prohibited Japan from immediately ratifying the protocol. First, the calculations assumed that carbon sequestration by all managed forests in Japan during the first commitment period (2008–2012) would be counted under Article 3.4 of the Kyoto Protocol. About 66% of land in Japan was covered by forest at that time; much of it had been planted in the 1950s and 1960s. Under Article 3.3, however, only afforestation, reforestation, and deforestation activities since 1990 could be incorporated into the calculation of emission reductions. To meet its target, Japan was depending on counting the sequestration of CO_2 by managed forests that had existed since before 1990, covered only by Article 3.4 of the Kyoto Protocol. Actually, Article 3.4 was the most critical article agreed upon at the last moment of COP3 for Japan, so it was viewed as an indispensable agenda item that needed to be fully utilized in Japan to achieve its emission reduction targets under the Kyoto Protocol.

The second assumption was that the Kyoto Mechanisms would be fully available to help Japan meet its target. Principles, modalities, rules, and guidelines for international emissions trading and the two other mechanisms were to be negotiated after COP3, and Japan could not be sure that it could rely on those mechanisms until those rules were agreed upon at the international level. Article 17 of the Kyoto Protocol stated that "any such trading shall be supplemental to domestic actions for the purpose of meeting quantified emission limitation and reduction commitments under the Kyoto Protocol." This sentence was inserted at the insistence of the EU, with the support of the developing countries, which asserted that emissions reduction should occur mainly at the domestic level. The criteria for judging a "supplemental" level had yet to be negotiated.

As a new phase of negotiations began in preparation for COP4 in 1998, Japan concentrated on making progress on what it considered to be two remaining contentious issues: the use of LULUCF and the use of the Kyoto Mechanisms. In addition to these two critical issues, there were many other agenda items remaining on the table in the international arena to be solved at COP4. In terms of compliance procedures, for example, Japan preferred a facilitative type of procedure rather than a punitive one. It was difficult for Japan to commit to an international emissions trading scheme if committing to binding consequences was a precondition for eligibility to participate in the Kyoto Mechanisms. Japan was particularly interested in the CDM because it was interested in strengthening political and economic ties with developing countries – especially those in the Asia-Pacific region.

The United States and some other non-EU countries also agreed that punishment should not be the consequence of not achieving emission reduction targets, although they considered some legal consequence would be needed for the Kyoto Mechanism to work smoothly. The EU and non–Annex I countries were more strict in the sense that some kind of punitive consequence for noncompliance was needed for the Annex I countries to sincerely aim at their respective emission objectives. These countries generally thought that, if there were no punishment for noncompliance, countries would not be sufficiently motivated to pay for the emission allowances or other measures required to meet their emission reduction targets.

As noted previously, energy consumption and CO_2 emissions in Japan had been increasing since 1990, but both plunged considerably in 1998, when CO_2 emissions decreased by about 3.5% from the previous year. This decline was, however, mainly a result of the stagnant economy, and most of the emissions reductions were observed in the industrial sector. Meanwhile, energy consumption was still increasing in the commercial and transportation sectors. This meant that CO_2 emissions were likely to resume increasing when the economy rebounded.

Japan was in agreement with other Umbrella Group member countries (former JUSCANZ countries plus Norway, Russia, and Ukraine) on various other issues. It advocated for no numerical limitations on the amount of transferable emission permits. Japan supported full fungibility (meaning that emission permits obtained by CDM projects, for example, could be sold to another country through emissions trading) and acceptance of forest-related projects under CDM.

COP4 was held in Buenos Aires in November 1998. At the start of the meeting, setting the agenda became a controversial debate. In the run-up to adopting the Kyoto Protocol, the United States, supported by Japan, had proposed the inclusion of the notion of "evolution" to set up another process after COP3 to discuss how some relatively wealthy non–Annex I countries could become Annex I countries. The COP4 agenda had reflected this and included an item on the agenda to discuss commitments by non–Annex I countries. Eventually, the parties agreed that the president of COP4 would consult the issue with relevant countries. The rest of the agenda was accepted.

The agenda of COP4 could be categorized into two groups. The first was a group of issues related to enhancement of UNFCCC. Articles 4.8 and 4.9 (articles dealing with countries that are vulnerable to adverse impacts of climate change and those that are vulnerable to adverse effects of the climate change mitigation actions taken by Annex I countries), finance, and technology transfer were some of the specific items where the developing countries sought to make meaningful progress. The second was a group of issues that, it was considered, still needed resolving regarding the Kyoto Protocol. These issues included creating detailed rules to make the Kyoto Mechanisms workable, specifying detailed rules for counting GHG emissions and sequestration by sinks, and defining compliance procedures. These were only some of the issues that needed to be resolved before the Kyoto Protocol could be ratified by the participating countries. Naturally, non–Annex I countries were more interested in the first group of issues, whereas Annex I countries were more interested in the second. As days passed with little progress, the non–Annex I countries' positions began to diversify. The oil-producing countries such as Saudi Arabia strongly insisted on not discussing detailed issues related to the Kyoto Protocol unless the same level of attention was paid to Articles 4.8 and 4.9, whereas countries in Latin America, Africa, and the AOSIS were interested in discussing rules to related to CDM.

Another important agenda item was the second review of the adequacy of UNFCCC Article 4.2(a)(b). The first review was conducted at COP1 and resulted in the adoption of the Berlin Mandate. Non–Annex I countries were opposed to starting discussion on their own emissions limitations because, in their view, Article 4.2's scope was only on the commitments of Annex I countries.

After a long debate during the ministerial-level negotiations, "The Buenos Aires Plan of Action" was adopted (UNFCCC 1998). The plan consisted of six pillars: (1) a financial mechanism, (2) development and transfer of technologies, (3) implementation of Articles 4.8 and 4.9 of the UNFCCC, (4) activities implemented jointly during the pilot phase, (5) the work program on the mechanisms of the Kyoto Protocol, and (6) preparation for the first session of the COP serving as the Meeting of the Parties to the Kyoto Protocol (CMP1), including work on the elements of the protocol related to compliance and on policies and measures for the mitigation of climate change.

Slow progress at COP5 and COP6

The year 1999 was not a fruitful one for the climate change dialog. The United States was still under the Clinton-Gore administration, which should have been favorable in terms of environmental issues, but President Clinton's scandal in 1998 had weakened his administrative power to push any political issues. Moreover, the Kyoto Protocol, which set absolute emission targets on Annex I countries, including the United States, was harshly criticized by industry- and energy-related stakeholders in the country as "totally flawed" (Cooper 2001). The US Congress showed no signs of taking up climate change as an agenda. In Japan, meanwhile, there was little intervention on existing climate change policies from the political side. Government officials continued to insist that the two same items, LULUCF and the Kyoto Mechanisms, were necessary for Japan to ratify the Kyoto Protocol.

COP5 was held in Bonn in December 1999. The outcomes of the Buenos Aires Plan of Action were to have been achieved in one year, but little progress had been observed. Developed countries were eager to discuss issues related to the rules for the Kyoto Mechanisms, compliance procedures, and sequestration by sinks, but developing countries – particularly oil-producing countries – insisted that no progress should be made in these agendas unless the same level of progress was observed in the agenda on Articles 4.8 and 4.9. In the ministerial session, countries expressed hope that the Kyoto Protocol would enter into force in 2002, the year in which the Rio+10 conference would be held. However, there were no particular reasons given to support this optimistic projection.

The Buenos Aires Plan of Action called for an agreement to be reached by COP6, but no agreement had been reached by the time it was held at The Hague in November 2000. COP6 President Jan Pronk circulated a draft of an agreement, but many issues remained unresolved on the final day (Grubb and Yamin 2001). The COP decided to suspend the session and reopen it in mid-2001.

The most serious agenda was related to rules concerning Article 3.4 of the Kyoto Protocol on LULUCF. The EU considered that emission mitigation should be the dominant effort of countries and preferred a limited use of sequestration by sinks. Countries that supported relatively full accounting of sinks, such as the United States, Canada, and Russia, generally were large and had vast amounts of forested areas. If these countries were able to fully count sinks, they would actually be able to increase emissions from GHG sources. The EU expressed concerns about the scale of sequestration, as well as of its permanence, because CO_2 sequestered by forests stays in forests only for several decades and then returns to the atmosphere when the trees decay. Meanwhile, the Umbrella group insisted that the use of sinks under Article 3.4 was a prerequisite for them to ratify the Kyoto Protocol; thus, for political reasons, the group would not compromise on this issue. The disagreement between the two sides could not be resolved, and this led to a decision to postpone the meeting. As a member of the Umbrella group, Japan also required the inclusion of sequestration by sinks to achieve its 6% reduction target. Although Japan was aware that a relatively weak rule on sinks would undermine the environmental effectiveness of the Protocol itself, Japan's priority was fixed on its own need to ratify the protocol.

The Kyoto Mechanisms remained a controversial agenda item at COP6. Article 17 of the Kyoto Protocol stated that "any such trading shall be supplemental to domestic actions for the purpose of meeting quantified emission limitation and reduction commitments." This reflected the EU's well-established position on this issue. Thus, the EU asserted that an upper limit on the acquisition of emission allowances and credits should be set for each country. The Umbrella group was opposed to this, contending that trading should be left to the market.

A series of disagreements were also observed in setting the rules for CDM. Japan wanted to include forest-related projects as well as those related to building nuclear power plants. Japan also hoped to utilize overseas development assistance to promote CDM, but this idea was opposed by the developing countries. Japan pushed for a simple procedure in defining the methodologies used to set baselines for CDM projects, but other countries held the view that the methodologies needed to be strictly defined. Article 12.8 of the Kyoto Protocol stated that "Parties to this Protocol shall ensure that a share of proceeds from certified project activities is used to cover administrative expenses as well as to assist developing countries that are particularly vulnerable to the adverse impact of climate change to meet the costs of adaptation." Some developing countries considered that a similar rule should be applied to the other two Kyoto Mechanisms, emissions trading and joint implementation, so as not to increase the price of certified CDM credits as compared to other types of emission allowances.

Compliance was another conflict-ridden agenda item. Countries were in general agreement that the noncompliance procedures for developing countries should be facilitative, not punitive, and include technical and informational support. Countries' views were more divided on noncompliance procedures for Annex I countries exceeding their emission reduction targets in the first commitment period. Japan argued that the consequence should not be punitive even for the developed countries, because some countries might choose to withdraw from the protocol rather than face the punishment.

In terms of finance, the developing countries wanted a mechanism that was independent of the Global Environmental Facility, an institution under the World Bank to which the UNFCCC entrusted its tentative financial mechanism. The group requested the establishment of new funds that focused on actions related to climate change only. Oil-producing countries again demanded progress on Articles 4.8 and 4.9, but the developed countries were willing to deal only with countries that had specific needs and concerns arising from the adverse effects of climate change, not with impacts of the implementation of response measures.

The United States' withdrawal from Kyoto Protocol

Early in 2001, the third assessment report of the IPCC was published. It claimed that the global average temperature was continuing to rise and that most of the rise could be attributed to the increasing atmospheric concentration of GHGs from anthropogenic emissions (IPCC 2001).

Also, in January 2001, the George W. Bush administration took power in the United States, which announced its withdrawal from the Kyoto Protocol in March. Prime Minister Yoshiro Mori was quick to send a letter to President Bush expressing his concern about the negative impact on international negotiations and urging the United States to continue working with Japan to aim at entering the Kyoto Protocol into force (Tiberghien and Schreurs 2007). The following month, the Japanese Diet unanimously adopted a resolution stating that Japan would stay committed to work toward entering the Kyoto Protocol into force and requesting that the United States return to the Kyoto Protocol regime. Environment Minister Yoriko Kawaguchi issued a formal statement, which highlighted Japan's intention to collaborate with the EU to ensure ratification of the Kyoto Protocol (Box 3.3).

Box 3.3   Statement on Climate Change by Yoriko Kawaguchi, Environment Minister (MOE 2001)

April 9, 2001

Today I talked with Mr. Kjell Larsson, the Swedish minister of the environment, and representatives from Belgium and the European Commission. It was a meaningful meeting in which we had productive discussions in a friendly atmosphere. At the beginning of the meeting, I expressed my appreciation for the active role that the European Commission has played in bringing progress to international negotiations through its visits to major countries.

Through our meeting today, my European colleagues and I confirmed the common views we share on the following points. First, Japan and the EU share concern about the pronouncement made by the U.S. President George W. Bush on not supporting the Kyoto Protocol. Second, according to the Third Assessment Report of IPCC, climate change is a matter of urgency and the developed countries have the responsibility of taking the lead in addressing it. Third, the Kyoto Protocol already contains tools that would enable cost-effective reduction of emissions and create economic opportunities. Fourth, Japan and the EU remain committed to their goal of realizing the entry-into-force of the Protocol no later than 2002, which would represent the fruit borne through no less than ten years of endeavors by the international community. Fifth, the U.S. participation is extremely important in ensuring the effective implementation of the Protocol and in strengthening global actions to combat climate change. Therefore, both Japan and the EU will continue to urge the U.S. Administration to reconsider its position on the Protocol. Finally, we agreed that Japan and the EU should keep close contact with each other.

I also informed my colleagues in the EU that the Japanese government is undertaking preparation for the establishment of a domestic regime that will enable it to fulfill its commitment made under the Protocol. Furthermore, I reported the following actions that Japan has taken recently with regard to the U.S. position. First, Prime Minister Yoshiro Mori sent a letter to U.S. President, Mr. Bush, and some ministers in the Japanese government, including myself, also wrote to our U.S. counterparts or otherwise made known our concerns. Second, a delegation made up of members of the Japanese ruling parties as well as senior representatives of the government visited Washington, D.C. last week to appeal to the U.S government and members of the Congress. Third, I held discussions with my colleagues from the Umbrella Group countries last week and we all agreed that it was important to continue working together as the Group. The U.S. was told that other countries saw the Kyoto Protocol as the basis of negotiations and that they looked forward to seeing the result of the policy review. Fourth, the environment ministers of Japan, South Korea, and China expressed, in a joint communique released yesterday here in Tokyo after their third tripartite meeting, their wish for the U.S. government to work actively with all Parties of the Framework Convention to attain a successful outcome of the resumed COP6 in order to bring the Kyoto Protocol into force as early as possible.

Japan reaffirms its commitment to continue urging, in collaboration with other countries, the U.S. Administration to reconsider its position on the Protocol and to actively contribute to negotiations so that the outcome of the resumed COP6 will allow the Kyoto Protocol to be ratified.

Although there were no major disagreements between the various ministries within Japan's government on the actual conditions required for early ratification, opinion was divided on the extent to which Japan should follow the lead of the United States. One group, especially those related to the industrial sector, thought that Japan should not ratify the protocol unless the United States did so. Some critics also viewed the Kyoto Protocol as unfair because they thought that emissions from some EU member countries had been reduced because of factors not directly related to climate, such as political and economic changes. With the enlargement of the EU, CO_2 emissions from the EU as a whole could be further reduced at a relatively low cost, especially if the base year remained 1990. A considerable segment of the Japanese industrial sector was unhappy with an approach that set the emission target relative to 1990 as the base year. They insisted that Japan had made energy efficiency improvements and energy substitutions in the 1970s in response to the two oil shocks in that decade. For Japan, it would be more beneficial to compare emission reductions relative to levels in the 1970s.

Another group felt that Japan should stop following the United States and start making its own way in seeking sound climate change policies, while at the same time continuing to urge the United States to return to the process. Because the Kyoto Protocol was the only international treaty at the time to bear the name of a Japanese city, people in Japan felt especially responsible for the protocol to enter into force as soon as possible. For the protocol to enter into force, major Annex I countries had to ratify it. Thus, Japan's position was to continue to aim for early ratification.

Sensing increasing divisions within Japan, the EU made greater concessions in closed meetings held before the reopened COP6 meeting (COP6-bis) to be held in Bonn in July 2001. Japan's terms on the volume of GHG emissions absorbed by forests were accepted to a large degree, and Japan's views on compliance measures were also generally accepted by the EU (Takahashi 2006). Acknowledging the major concessions made by the EU, Prime Minister Koizumi decided that Japan should put aside the question of how to respond to the United States withdrawal temporarily and accepted the agreement. The G8 Summit was held in July, just before the COP6-bis, in Genoa, Italy, and Prime Minister Koizumi advocated for the need for the Kyoto Protocol's entry into force at the summit.

The Japanese government's position was clear by the time COP 6-bis opened in July. As was previously agreed upon by the EU, Japan's terms on the volume of GHG emissions absorbed by forests were accepted to a large extent, and its views on compliance measures were also reflected in the final text. The agreement reached in this meeting was called "the Bonn Agreements on the Implementation of the Buenos Aires Action Plan" (UNFCCC 2001a). In the document, Japan was permitted to count LULUCF sinks of as much as 13.00 MtC per year during the first commitment period. This was more than sufficient to fulfill the 3.7% sequestration target set by the *Guideline of Measures to Prevent Global Warming* in June 1998. In this sense, the withdrawal of the United States from the Kyoto Protocol worked to benefit Japan with regard to the stringency of Japan's commitments under Kyoto Protocol.

The Bonn agreement was reorganized and agreed upon at COP7, held in October–November 2001 in Marrakesh. Although there was some reopening of negotiations – for example, to work out Russia's demand for increased upper limits for counting sinks – the discussions went smoothly and resulted in the "Marrakesh Accord" (UNFCCC 2001b), which was more than 240 pages long. The parties agreed on modalities, rules, and guidelines for the Kyoto Mechanisms. New terms were defined to distinguish different types of tradable emission rights: the "assigned amount unit" was a unit issued pursuant to the relevant provision concerning emissions trading, the "emission reduction unit" was a unit issued for joint implementation, the "certified emission reduction unit" was a unit issued pursuant to CDM, and the "removable unit" was a unit issued pursuant to LULUCF-related sequestration. All four types of unit were considered tradable.

Under emissions trading, a loose upper limit of the tradable amount was set as follows: "each Party included in Annex I shall maintain, in its national registry, a commitment period reserve which should not drop below 90% of the Party's assigned amount, or 100% of five times its most recently reviewed inventory, whichever is lowest." In addition, an executive board was established to supervise and give guidance to the CMP regarding CDM, and the share of proceeds was accredited only under CDM.

The Compliance Committee was established; two branches – namely, the facilitative branch and the enforcement branch – deal with matters related to compliance. The mandate of the facilitative branch was to watch all countries' activities in general set out in the Kyoto Protocol. The mandate of the enforcement branch was to be responsible for determining whether each Annex I country was compliant with the terms for meeting its QELROs, its use of methodological and reporting requirements, and its eligibility requirements for the Kyoto Mechanisms. If an Annex I country was judged as not meeting its QELROs, it would face three types of consequences: (1) deduction from the party's assigned emission amount for the second commitment period at a rate of equal to 1.3 times the amount (in tons) of excess emissions; (2) development of a compliance action plan; and (3) suspension of eligibility to sell emission units via the emissions trading scheme.

Funding was another important pillar of the agreement. Three new funds were established under the Marrakesh Accords. The Special Climate Change Fund and the Least Developed Countries Fund were set up under the UNFCCC, whereas the Adaptation Fund was agreed upon under the Kyoto Protocol.

The Marrakesh Accords officially allowed Japan to count sequestration by managed forests, as defined in Article 3.4, for the first commitment period. It also established the rules necessary to get international emissions trading and CDM started. Japan considered this agreement to be satisfactory and began the domestic procedure towards ratification of the Kyoto Protocol. The Global Warming Prevention Headquarters revised its *Guideline of Measures to Prevent Global Warming* in March 2002 to adjust domestic policies to reflect the agreements made in the Marrakesh Accords (Box 3.4).

Box 3.4   Revised Guideline of Measures to Prevent Global Warming (Global Warming Prevention Headquarters 2002)

- CO_2 emissions from energy sources shall be the same level as that of fiscal year 1990 during 2008 to 2012 (see Table 3.1 for planned measures to achieve the goal).
- CO_2 emissions from non-energy sources, CH_4, and N_2O, shall be reduced 0.5% from the fiscal year 1990 levels during 2008 to 2012.
- An additional 2% reduction shall be achieved by using innovative technology.
- Growth in HFCs, PFCs, and SF_6 emissions should be limited to 2% relative to total GHG emissions of the base year between 1995 and 2008 to 2012.
- Removal of CO_2 by sinks stipulated in Articles 3.3 and 3.4 shall be about 3.9% of emissions in 1990.
- The remaining gap between the 6% reduction target and domestic emissions (1.6%) is to be eliminated using the Kyoto Mechanisms.

Table 3.1 Measures in the guideline to stabilize CO_2 emissions from energy sources

	Industry sector	*Residential and commercial sectors*	*Transportation sector*
Energy conservation: additional measures for reduction of 22 million tons of CO_2	* Voluntary action plans and their follow-up (the Keidanren Voluntary Action Plan on the Environment aims at a reduction to below plus minus 0%) * Technological development and diffusion of high-performance boilers and lasers * Promotion of the introduction of high-performance industrial furnaces	* Introduction of energy management systems for large office buildings in line with those in place in large factories through revision of the *Law Concerning the Rational Use of Energy* * Application of Top Runner Approach to gas equipment, which hitherto have not been subject to the standard, to further expand their use	* Accelerated introduction of vehicles conforming to the Top Runner Approach, accelerated development and diffusion of low-emission vehicles including clean energy vehicles * Traffic flow measures such as promotion of Intelligent Transport Systems (ITS) * Improvement of efficiency in freight services through means such as a modal shift to marine transport

	Industry sector	Residential and commercial sectors	Transportation sector
		* Promotion of the diffusion of high-efficiency water heating appliances * Promotion of the diffusion of energy management systems for households and businesses	* Promotion of the use of public transportation
New energy: additional measures for reduction of 34 million tons of CO_2	* Inclusion of biomass and snow/ice in the *Law Concerning Promotion of the Use of New Energy* * Proposal for enactment of *Law Concerning Promotion of the Use of New Energy by Electrical Utilities* * Promotion of assistance for the introduction of solar energy generation, solar thermal utilization, wind power generation, waste power generation, and biomass energy * Strengthening of technological development and practical testing of fuel cells, solar energy generation, and biomass energy		
Fuel conversion: additional measures for reduction of 18 million tons of CO_2	* Support for conversion of aging coal-fired power generation to natural gas power generation * Support for fuel conversion in industrial boilers * Development of safety standards for natural gas pipelines		
Promotion of nuclear power	* Promotion of nuclear power generation with the major premise of ensuring safety * Promotion of measures for development in regions of electric power generation in relation to the location of nuclear fuel cycle facilities		
	Approximately 462 million tons of CO_2 (+7%)	Approximately 260 million tons of CO_2 (+2%)	Approximately 250 million tons of CO_2 (+17%)

* The figures in parentheses are the percentage of reduction in emissions from FY1990 levels in each sector.
* Emission reduction targets for each sector are set as estimated targets that can be achieved on the basis of various premises and conditions.
* The assessment of measures should be carried out with a certain degree of flexibility and a perspective of the entire energy supply and demand structure.
* As the use of the Kyoto Mechanism by businesses is approved, it is expected that they will use the mechanism to achieve their reduction in a more cost-effective way.

Source: Excerpt from Global Warming Prevention Headquarters (2002)

Interestingly, the preamble paragraphs of the guideline were substantially revised from the 1998 version. In the initial version drafted in June 1998, the guideline started with an introductory section called the Basic Approach, which explained the seriousness of climate change and highlighted three points considered to be fundamental to the thoughts underpinning the guideline: (1) promotion and application of all conceivable innovative technologies; (2) promotion of measures to prevent global warming by encouraging private citizens to review their own lifestyles; and (3) development of international frameworks for measures such as emissions trading to enhance global action. The revised guideline also had an introductory section called Basic Principles, but the context was tilted more toward economic development: (1) create a balance between the environment and the economy to link the efforts to prevent global warming to economic revitalization and employment creation; (2) undertake assessment and review of the progress of measures being taken at regular intervals (in 2004 and 2007) and take necessary measures in a step-by-step fashion; (3) promote the combined effort by all sectors of society (meaning that the government would continue to promote voluntary initiatives by businesses while also strongly advancing the agreed-upon measures, particularly in the residential, commercial, and transportation sectors); and (4) ensure international cooperation for all measures by continuing to make the utmost effort to establish a common regime in which all countries, including the United States and developing countries, participate. Points (2) and (3) in the revised guideline reflected the government's cautious position that implementation of emission mitigation policies would be limited to a minimum level and that any additional policies would be implemented only after future revision of the current policies.

After the guideline was revised, the *Law Concerning the Promotion of the Measures to Cope With Global Warming* was also amended. In March 2002, MOE and MOFA drafted the amendment to the law. Although the first Koizumi Administration was a coalition administration consisting of the Liberal Democratic Party (LDP), Komei Party, and New Conservative Party, the New Conservative Party opposed the draft amendment. The New Conservative Party had gained the support of industry groups, and it put forth several conditions for them to accept the amendment. The first condition was that the domestic climate policies would be implemented only after it was sure that the Kyoto Protocol had entered into force through the participation of the United States. It was a sign that Japanese industries were starting to retreat from the Kyoto Protocol's 6% reduction target without US participation. The previous year, the Central Environment Council had established the Expert Committee on a Tax System to Combat Climate Change in October 2001, which recommended the "greening" of Japan's tax system. Industries had criticized the implementation of a carbon tax after these recommendations were made, and the distrust between MOE and industries affected the New Conservative Party's position on climate change policy. The second condition stressed that the procedure agreed in the Marrakesh Accord for dealing with non-compliance in meeting emission reduction targets should be neglected, because Japan can withdraw from Kyoto Protocol at any occasion. Last but not least, they wanted the Keidanren Voluntary Action Plan to be used as the main guideline for policies and measures for emission mitigation in the industrial sector. These conditions were criticized severely by the Democratic Party of Japan (DPJ), and the law was amended without meeting all the conditions raised by the New Conservative Party (Watanabe 2011: 58). In a way, this could be interpreted as a success, because it prevented the institutionalization of voluntary approaches. On the other hand, it became more difficult to implement a carbon tax at this stage, as it was considered important to improve relationship between MOE and industries.

Japan ratified the Kyoto Protocol on 4 June 2002. Even with all these new developments, however, emissions were still on the rise in Japan, especially in the residential/commercial and transportation sectors. It was clear that additional measures were needed to achieve the 6% reduction target, but the government's position was to "wait and see" if the Kyoto Protocol really would enter into force.

The World Summit on Sustainable Development (WSSD) was held in Johannesburg from late August to early September in 2002. This was the tenth anniversary of the Rio Earth Summit in 1992. Prime Minister Koizumi attended the meeting. Yoriko Kawaguchi had been appointed Foreign Affairs Minister in February 2002, and she also attended the WSSD with Environment Minister Hiroshi Ohki, who was president at COP3. Koizumi participated in a round-table discussion and made a speech in which he addressed his "Koizumi Initiative." The initiative consisted of three elements:

1   People and hope: people are the key factor for sustainable development, and education is of paramount importance. Japan promised to provide no less than JPY250 billion (USD 2.1 billion) in education assistance over a five-year period to the least developing countries. Also, health is just as essential as education.
2   Ownership and partnership: development of respect for the ownership of the recipient and extending support as an equal partner. Trade is an opportunity to assist the least-developed countries (LDCs), for example, by establishing duty-free and quota-free treatment of LDC products. Japan promised emergency food aid of USD30 million to countries in southern Africa.
3   Today's complacency, tomorrow's plight: entry into force of the Kyoto Protocol is a necessity and a common rule for all. Japan pledged assistance in environment-related capacity-building by training 5,000 people from overseas over a five-year period.

Unlike in 1992, when the prime minister did not attend, Japan was able to demonstrate a positive attitude toward the global discussion of sustainability at this meeting.

Political leaders through COP3

This chapter began with a simple question – why did Japan decide to host COP3? On some occasions, political leaders have taken the lead in affirmative climate policy actions (Schreurs 1997). For example, some leaders of LDP were influential in stimulating the government toward taking stronger action during the climate change debate. This was, however, effective only when the various ministries had unresolvable conflicting ideas that had to be settled by outsiders. Meanwhile, an ongoing lack of political leadership during closed-door policy-making meetings among ministries and agencies was observed during internal negotiations on Japan's proposal for the emission targets in the Kyoto Protocol.

After the 1991 Gulf War, Japanese leaders were able to pass legislation that enabled Japanese military troops to participate in the United Nations Peace Keeping Operations (PKO). Ironically, however, pressure from the United States for Japan to contribute to the PKO withered after the failure of the mission in Somalia in 1993–1995. Nevertheless, the intention of Japan's diplomacy was to contribute to international activities to win a permanent seat on the United Nations Security Council (UNSC). Thus, Japan continued generously to support the UN system financially, while also maintaining bilateral relationships with key partners. Japan's rationale for its inclusion in the UNSC is regarded as an element of a wider reform of the UN system aimed at reflecting the realities and power distributions of the post–Cold War world, rather than those of the post–World War II world (Hook et al. 2012: 325–327). A permanent seat on the UNSC was targeted continuously by MOFA.

The five-year period between the adoption of UNFCCC in June 1992 and the adoption of the Kyoto Protocol in 1997 was a time of political unrest in Japan. The prime minister at the time of COP1 was Tomiichi Murayama, the first political leader from a non-LDP party since the LDP began governing in 1955. Murayama was head of a coalition administration with the LDP, the Social Democratic Party of Japan (SDPJ), and a newly established party called Sakigake. Tomiichi Murayama was one of the SDPJ's leaders; he took office in June 1994 and served through January 1996. The LDP and SDPJ had been long-time political opponents. For the SDPJ to "shake hands" with the LDP (i.e., form a coalition), it had to abandon two important principles that it had long espoused: opposition to sending the Japanese military abroad and opposition to a security treaty with the United States. The Murayama administration also coincided with the widening of the diplomatic gap between Japan and the United States. The Japan-US trade talks over automobiles, rice, and other agricultural products resulted in serious controversy in both countries. This economic conflict prompted Japan to take countervailing power against the United States by impressing the world with its leadership in global environmental issues (Matsumura 2000).

Murayama attended the G7 Summit in Naples in July 1994 just one week after his inauguration. This circumstance made it difficult for Japan to negotiate anything of substance at the summit. In addition, in early 1995, the Murayama Administration had to deal with two serious incidents that struck Japan, reducing the time and political will needed for debates on climate change or many other important issues. The first was a magnitude 7.2 earthquake that occurred on 17 January 1995 in the Hanshin area, leading to more than 6,000 deaths and widespread property destruction. The cost of the damage was estimated to be

JPY9,600 billion. Murayama and many other political leaders initially did not realize the seriousness of the earthquake because so much of the infrastructure, including telephone lines, was disrupted. Murayama was criticized for his late response.

On 20 March of the same year, a sarin gas attack occurred on the Tokyo subway system in the heart of Tokyo. Thirteen people were killed, and more than 6,000 people were injured. Aum Shinrikyo, a newly established religious group, was suspected of involvement in the attack, and the government quickly initiated an investigation of the Aum Shinrikyo's headquarters in Yamanashi Prefecture. These incidents incited fear throughout Japan. Faced with these historic events, political leaders had little time to worry about environmental issues. In a vacuum of political leadership, MOFA and the EA pushed their agenda to have Japan host COP3.

Japan's relationship with other northeast Asian countries was also changing during this period. During the administrations of Morihiro Hosokawa (the previous prime minister) and Murayama, increasingly explicit apologies were made for Japan's past actions in northeast Asia. Historically, since the end of World War II, Japan tried to avoid any direct discussions of the history of the relationships between Japan and China and Korea. This was a crucial burden on Japanese foreign policy in the region. Murayama and the SDPJ were more willing than LDP politicians to make an official apology and move forward. Murayama remained committed to improving relationships with China and Korea. The fiftieth anniversary of the end of World War II was marked in 1995, and Murayama considered it important to make an announcement as part of a formal dialog. What became known as the Murayama Danwa (discourse) was published on 15 August 1995. In the dialog, he stated,

> during a certain period in the not too distant past, Japan, following a mistaken national policy, advanced along the road to war, only to ensnare the Japanese people in a fateful crisis, and, through its colonial rule and aggression, caused tremendous damage and suffering to the people of many countries, particularly to those of Asian nations. In the hope that no such mistake be made in the future, I regard, in a spirit of humility, these irrefutable facts of history, and express here once again my feelings of deep remorse and state my heartfelt apology. Allow me also to express my feelings of profound mourning for all victims, both at home and abroad, of that history.
>
> (MOFA 1995)

The term "apology" was used to express the feelings of Murayama as an individual and not as the dead of Japan as a nation-state, but this dialog has been perceived as coming the closest of any statements made by Japanese prime ministers to being a formal apology to Japan's neighboring countries. This issue reemerges every decade, even in today's regional affairs. The relationship with China and Korea clearly dominated foreign policy during the Murayama administration.

The negotiating process leading up to the Kyoto conference was mostly synchronized with Prime Minister Ryutaro Hashimoto's administration between January 1996 and July 1998. Despite the relative length of that administration (from the Japanese perspective), from which climate policy could have benefited, the

Hashimoto Administration was politically unstable throughout its duration. The administration was still made up of a coalition of the LDP, SDPJ, and Sakigake, indicating the continued relatively weak status of the LDP within the administration. Hashimoto was also engaged in raising the value-added-tax rate from 3% to 5%; this went into effect in April 1997. The timing of the tax increase was not the best, because Japan's GDP growth had decreased more than predicted. Immediately after the adoption of the Kyoto Protocol at COP3 in January 1998, scandals dominated the media. Many of the major banks and stock exchange companies in Japan had been hosting expensive dinners for MOF officials to gain favor.

Despite these political disadvantages, Ryutaro Hashimoto himself was committed to global environmental issues and working on climate change. This was partially due to his background in the field of social welfare. During the time when Hashimoto was minister of finance, he considered the implementation of an environment tax and instructed government officials to conduct a study to see if investments in environmental protection measures could coexist with economic growth. The first environment minister appointed by Hashimoto, Sukio Iwatare, was the only one in the past three decades that belonged to the SDPJ. He was considered "green," so he was happy to support the LDP in hosting COP3 in Kyoto. He had attended the COP2 meeting; he thought that Japan should show leadership in tackling global environmental problems and supported the Geneva Ministerial Declaration. The next environment minister, Michiko Ishii, however, was trained as a chemist and had little political power to influence governmental decisions on emissions reduction targets in the first commitment period of the Kyoto Protocol. As explained in the previous sections, the bureaucrats had serious conflicts among themselves in the summer and fall of 1997, but there was little, if any, intervention by political leaders, including Ishii, regarding Japan's emission reduction target. As negotiations entered the point where countries had to determine their emissions reduction targets, MITI, with full technical support of industry and business groups, became a powerful agency advocating for less ambitious emission reduction targets. In September 1997, Deputy Chief Cabinet Secretary Teijiro Furukawa, an able bureaucrat and not a politician, took the role of mediator among MITI, EA, and MOFA.

While the government officials in relevant ministries were engaged in heated debates over Japan's emission reduction target, the cabinet members changed, and Michiko Ishii resigned in September 1997. A new environment minister, Hiroshi Ohki, was appointed just before COP3 in October 1997. One of the reasons why Ohki may have been appointed was that he spoke English. At that time, many senior Diet members did not speak English, and it was necessary for the EA to have someone who could speak English to be the president of COP3. This rapid change of environment ministers did not help Japan in taking initiative in the AGBM negotiations. At the same time, Prime Minister Hashimoto became more engaged in the final round of negotiations. Japan had long relied on the United States for its national security; the US had also been one of Japan's largest trading partners, even though American industries were competitors with Japanese industries. For many Japanese bureaucrats and politicians, maintaining good relations with the United States was considered to be the safest way to secure prosperity (Hasegawa 2004). Hashimoto cherished good bilateral relations with the United States. As a typical example, he was eager to reconfirm the US-Japan security treaty by revising the 1978 *Guidelines for U.S.–Japan Defense Cooperation* in 1996. Hashimoto and President Clinton agreed to relocate a US military base from Futenma, Okinawa, to Japan, and to build a new helicopter base in Henoko, Okinawa, although this arrangement has yet to be implemented because of local opposition in Okinawa.

Thus, during the negotiations leading up to COP3, the Japanese government became more and more inclined toward satisfying the conditions laid out by the United States to accept the Kyoto Protocol. When the United States accepted a 7% reduction target, Japan almost had no other choice but to accept a 6% target for itself. Although Hiroshi Ohki was the president of COP3, and thus could have used his position to benefit Japanese government, he intentionally acted with neutrality and stayed away from internal inter-ministerial conflicts (Ohki 2007). Hashimoto acted immediately after adoption of the Kyoto Protocol to establish the Global Warming Prevention Headquarters in the cabinet only ten days after COP3, leading to enactment of the *Law Concerning the Promotion of the Measures to Cope With Global Warming* in 1998.

Another important role Hashimoto played during his administration was restructuring of the government. There were many problems within the central government, including budgetary constraints. The entire government needed to be downsized, but none of the ministries wanted to experience cuts. It was thus the role of the politicians to conduct the work. Whereas most of the ministries were downsized, only the Environment Agency was elevated to become the Ministry of the Environment, with enlarged budgetary and human resources.

Political leaders after COP3

Keizo Obuchi served as prime minister from July 1998 to April 2000. Obuchi was interested in using "human security" to show Japan's contribution as a middle power in the international arena. He declared that the twenty-first century should be a "human-centered century," and he announced that under the United Nations system Japan would establish a Human Security Fund, to which Japan offered more than USD175 million between 2001 and 2003. He also invited G8 members to a summit in Okinawa, which was to be held in July 2000, and he wanted the issue of human security to be one of the main agenda items at the meeting. Unfortunately, he was hospitalized in spring 2000 with a cerebral infarction and had to step down from his post. He passed away in May of the same year. Because Obuchi was the foreign affairs minister at the time of COP3, where he made some opening remarks, he ought to have had some level of understanding of the Kyoto Protocol and climate change negotiations. Circumstances, however, did not allow him the time to respond more positively to the climate change agenda. MOFA officials also placed greater emphasis on human security than on climate change during the Obuchi administration.

Yoshiro Mori was appointed prime minister immediately after Obuchi's resignation in April 2000 and remained in power through April 2001. No evidence could be found that he had an interest in global environmental issues. One of the contentious foreign policy issues of the time was Japan's relations with North Korea and Russia. Mori was able to develop a good relationship with Russia's president, Vladimir Putin, but he was unable to make any progress with North Korea, at least in part because of his own political weakness internally and because of growing pressure within Japan over the abduction issue (in which North Korea was accused of abducting Japanese citizens). In addition, the United States withdrew from the Kyoto Protocol only a month before Mori resigned, and the LDP was already having internal discussions on his replacement.

A new Environment Minister, Yoriko Kawaguchi, was appointed in August 2000. Her career began as a MITI official, and she had accumulated the knowledge and skill needed to negotiate the detailed rules in the climate change negotiations leading up to the Marrakesh Accords. She was reappointed as environment minister even after Junichi Koizumi became prime minister.

Koizumi became prime minister in April 2001, after winning an election internally within the LDP to choose the head of the party. He emphasized the need for changes in both the domestic and diplomatic arenas (Hook et al. 2012). His aims were to downsize the role of the central government, privatize government-run organizations as far as possible, reduce industry regulations, and shrink national expenditures to reduce the deficit. He increased the number of staff in the cabinet to support his duties. Previously, three offices of the assistant chief cabinet secretariat were installed under the cabinet, and the staffs were obtained from three different ministries, mainly from MOF and MITI. This structure was useful for reflecting bureaucratic ministerial interests in the decision-making of political leaders. The reform merged these positions into one, and that office was led by staff sent from the various ministries. This reform resulted in the cabinet having a stronger influence when dealing with complex domestic and foreign policies.

One of the targets of change was the privatization of the Japan Highway Public Corporation, a fully government-funded organization that promoted construction of toll roads across Japan. Its funding was collected from taxes on gasoline and payments by highway users. Politicians supporting this organization belonged to the road-*zoku* and were well-known for the vast amounts of political power to make decisions on budget allocation. Koizumi intended to remove the Japan Highway Public Corporation from the national government and have it conduct business independent of the government. There was firm opposition from the road construction industries and also from local communities hoping to have highways constructed in their areas, but the laws to privatize the corporation were finally adopted in 2004. The process of privatizing the Highway Corporation was strongly connected to discussions on a carbon tax, because a gasoline tax already existed, but the proceeds were used for purposes unrelated to climate policies. Environmentalists hoped that, with the privatization of the Highway Corporation, the tax collected could be spent on climate policies, but the tax revenue was under the authority of the Ministry of Economy, Trade, and Industry (METI, formerly MITI). METI's intention was to reform the entire energy tax system and use the revenue to support the electricity industry – particularly the nuclear power sector.

Article 68 of the Japanese Constitution stipulates that more than half of the members of the cabinet must be chosen from the Diet, but, in practice, cabinet members were almost always chosen from the Diet. Koizumi, however, appointed three people from the business community and academia as ministers because he wanted real experts in the cabinet. He also appointed five female ministers; this was the largest share of women in the post–World War II history of the cabinet. Yoriko Kawaguchi, one of the female ministers with a nonpolitical background, was reappointed as environment minister through February 2002, when she became the foreign affairs minister.

Although Koizumi was eager to decrease the worsening government budgetary deficit, he also was aiming to create a smaller government. Therefore, he was not interested in raising the consumption tax rate, which had been an agenda item for most of the prime ministers in Japan in the previous 20 years. This opposition did not create favorable conditions for those who wished to introduce an environmental or carbon tax.

In the foreign policy arena, Koizumi, similarly to most previous LDP prime ministers, continued to foster good relations with the United States. Soon after the attack in New York on 9/11, Koizumi sent clear messages to President Bush that Japan supported the United States in the fight against terrorism. In addition, soon after the attacks, the Japanese government offered USD10 million in assistance to the families affected by the attacks. Japan also sent troops to the war in Iraq to work in the field of reconstruction of the region.

Koizumi was proactive in his attempts to improve relations with North Korea. It came as a great surprise, even for many LDP members, when Koizumi announced in August 2002 his plan to go to North Korea to meet President Kim Jong Il the following month to discuss security issues and the abduction of Japanese people by the North Korean government. The meeting was a success. Under the agreed upon Pyongyang Declaration, North Korea affirmed its commitment to freezing its nuclear program and its moratorium on missile testing. It also admitted to involvement in the abduction of 13 Japanese citizens, eight of whom had since died in North Korea (Hook et al. 2012: 198). The survivors were permitted to visit Japan within a month of the meeting. The two countries also had an ongoing dialog on working toward normalization of their relationship, but meager progress was made. At about this time, the major forum for interaction between the two countries shifted to the Six Party Talks in Beijing, which included China, South Korea, Russia, and the United States, in addition to Japan and North Korea.

In the context of these evolving foreign policy circumstances and his close relationship with President Bush, Koizumi might have had a different response if the US withdrawal from the Kyoto Protocol had occurred after his inauguration.

Involvement of nonstate actors

Emergence of environmental NGOs

At the time of COP1, one of the goals of the EA in hosting COP3 was to stimulate Japanese citizen groups to be more aware of environmental problems. This aim certainly appears to have been fulfilled.

Traditionally, Japan had little experience in nurturing volunteers or supporting NGO-related activities. A turning point occurred in January 1995, just before the COP1 meeting, when the Hanshin-Awaji Earthquake struck the central region of the Japanese archipelago. More than 20,000 individuals were said to work as volunteers in the hard-hit area to help the residents. The support of these volunteers was said to exceed that of the national government. After the earthquake, many citizens groups called for institutional, legal support of NGOs. One way to support them was to introduce tax exemptions on donations to certified NGOs. The first legislation to support NGOs – the so-called Non-Profit Organization Bill – was enacted in 1998.

The decision to host COP3 in Kyoto stimulated a tremendous level of public interest in Japan. To prepare for the upcoming Kyoto meeting, environmental NGOs that were already in existence gathered to establish a network called the Kiko (Climate) Forum in December 1996, a year before COP3. This was the first NGO in Japan to deal mainly with the climate change problem. The Kiko Forum later developed into the Kiko Network in April 1998, and it has become the most influential environmental NGO in Japan on climate change policy.

Another NGO, the Green Energy Network, was established in May 1999 with the aim of increasing the diffusion of local renewable energy in Japan. The group proposed a new bill to introduce a feed-in tariff – a system that would assure that electricity produced from renewable energy sources would be purchased by power companies at higher prices than electricity generated by thermal power plants. The Institute for Sustainable Energy Policies, another NGO aiming at the diffusion of sustainable energy resources in Japan, was established in September 2000.

These NGOs share a common feature. Traditionally, most Japanese citizen groups on environmental issues were either pressure groups that insisted on environmental conservation or gatherings of people who shared a hobby, such as bird watching or mountain climbing. The newly emerged NGOs' objective was to change the status quo by being engaged in decision-making in the government's energy policies, which had been conducted behind closed doors up to this point. The NGOs had built enough capacity to deliver high-quality analyses and policy recommendations. They also had the skills to communicate in English and collaborate with NGOs outside Japan, such as the Climate Action Network, to be able to pressure the Japanese government both domestically and from abroad.

Industry

Keidanren started internal discussions on how to respond to the climate change debate in the mid-1990s. The Keidanren Voluntary Action Plan was published in

June 1997. As noted previously, many of the targets were not absolute emission reduction targets but relative targets, such as emissions per unit of production (Keidanren 1997). With the announcement of the plan, Keidanren also stated that they would aim to voluntarily stabilize CO_2 emissions from the industrial sector to 1990 levels by 2010. From one perspective, this action plan appeared as though the industries were being cooperative with the government on climate change mitigation, but it also represented a trade-off in that they wanted to pressure the government to not demand further reductions. In September 1997, the Keidanren made an appeal to the government, noting that industry had already implemented a great deal of energy-saving technology since the oil crises in the 1970s, and that the emission reduction targets should fully account for such past efforts. They were also clearly supportive of nuclear energy and opposed to the introduction of a carbon tax or any other environmental taxes.

The actions of these industries have been reviewed every year since the adoption of the Kyoto Protocol. In large part because of economic stagnation, industrial energy consumption did not grow as much as was projected. Nevertheless, the 6% emission reduction target was perceived as being quite difficult to achieve, especially by energy-intensive industries.

The Keizai Doyukai, another group consisting of leaders of the business community, also responded to the climate change problem, but in a slightly more positive manner than the Keidanren and with clear reference to the need for nuclear power. In November 1997, just before COP3, the group made an announcement on their five pillars for tackling the climate change issue: (1) each individual person needs to consider climate change as his or her own problem and change his or her lifestyle; (2) industries and businesses should act in advance of other stakeholders; (3) environmental costs should be internalized in the market economy; (4) nuclear power should be promoted vigorously; and (5) the development of innovative technology was indispensable.

National emissions continued to increase through 2000, and there were no sufficient policies or measures in place in Japan to change the trend. Among the Japanese industry groups, the Keidanren strongly opposed both a carbon tax and emissions trading, insisting that emissions from the industrial sector had been stabilized already and that such policies would harm the economy. The group noted that leading Japanese industries would lose competitiveness under the current regime because industries in the United States and developing countries would not face the same constraints. In addition, after the United States withdrew from the Kyoto Protocol in 2001, opponents of the Kyoto Protocol called it a failure because it allowed the largest emitter in the world to so easily abandon its commitment. Japanese industries were almost unanimous in their criticism of the Kyoto Protocol, including on opposing the introduction of a carbon tax and emissions trading and the desire to push voluntary agreements.

Scientific organizations and researchers

The number of Japanese scientists dealing with climate change increased in the 1990s, and increased research funding was allocated to the theme. Various scientists reported on the latest findings on the potential impacts of climate change in Japan in the future (Harasawa and Nishioka 1997). There was, however, still little direct connection between Japanese scientists and government decision-making. There was almost no debate on whether climate change science was too uncertain to take any action. People generally accepted the IPCC's work and trusted the claim that climate change caused by anthropogenic GHG emissions was causing problems. The main theme of the debate was, therefore, by how much Japan should reduce GHG emissions, in relative terms, as compared with other major economies.

The AIM modeling team was one of a few expert groups that were heavily involved in negotiations between ministries. The model was developed in 1990 under the leadership of Tsuneyuki Morita of the NIES, as a response to similar modeling projects that were already in progress in the United States and Europe, so that Japan and its scientists could participate in the IPCC Working Group 3 drafting work. The model was recognized internationally, but the modeling team faced harsh criticism, which came out from political and economic stakeholders, rather than technical or academic-oriented individuals. It was unfortunate that Japanese decision-makers were not prepared for a scientifically neutral debate at the time of COP3.

Local government

Japan's local governments had not been engaged in global environmental issues before COP3. Local authorities generally considered local pollution (since the 1960s) and newer environmental agendas such as waste management as being within the scope of their authority. However, the rising importance of environmental NGOs, together with local political willingness to become more independent from central bureaucratic and political authorities, pushed local governments to work independently toward greener regional development (Kameyama 2002). These movements were put into practice as local emission reduction targets and the development of renewable energy as a regional source of energy.

The *Law Concerning the Promotion of Measures to Cope With Global Warming* obliged local governments, mainly prefectures, to develop action plans to mitigate climate change. This law incentivized local governments all over Japan to create their own individual ideas to deal with climate change mitigation. Within only a few years of the law's enactment (by about 2000), more than 100 local governments (municipalities and villages) initiated activities to develop renewable energy, such as wind and solar power, under their own authority. Even during the 1950s and the 1960s during the local pollution crisis, local governments generally took action before the national government was able to respond fully to the serious health hazard posed by pollutants. Once again, the local communities were the first to engage in mitigating GHG emissions within their boundaries.

The headquarters of the International Council for Local Environmental Initiatives (ICLEI, renamed Local Governments for Sustainability in 2003) was established in 1990, and ICLEI Japan was set up in 1993. Major megacities in Japan, including Tokyo and Kyoto, have become members of the organization.

Members of the public

In the late 1980s, as Japanese citizens began to gain awareness of global environmental problems, many began to view climate change as a serious problem. A public opinion poll conducted by the Prime Minister's Office in June 1997 indicated that a majority of the people were interested in the global warming problem (very interested, 25.3%; interested, 54.1%; not that interested, 15.8%; and not interested at all, 3.9%). Most people were aware that they could make a contribution toward mitigating climate change, and many had already taken some action to save energy. Only 8.8% of the respondents said they were taking no action to mitigate climate change (Prime Minister's Office 1997).

Nevertheless, people's willingness to contribute toward GHG emission mitigation was somewhat reduced after a serious nuclear power accident occurred at Tokaimura in September 1999. Three months before the accident, the Agency for Natural Resources and Energy had set up a committee to revise the Nuclear Power Long-Term Plan. The committee was expected to finalize the report in the direction of expanding the use of nuclear power plants, partly gaining official support because of the Kyoto Protocol's 6% reduction target. The report indicated that 20 more nuclear power plants would be needed to achieve the target. The accidental release of radioactive materials at the fuel fabrication plant in Tokaimura resulted in two deaths, and more than 600 people were affected by radiation. This incident drew people's attention to the risks associated with nuclear power, and, as a consequence, people's support for climate change mitigation, at least through the use of nuclear power, was diminished.

Summary

This chapter has given an overview of the period between 1995 and 2002, during which Japan negotiated the Kyoto Protocol, hosted COP3, and finally ratified the Kyoto Protocol. Japan's initial motivation for hosting COP3 was to highlight its contribution to international affairs and secure a permanent seat on the UNSC. Although Japan did host COP3 and the Kyoto Protocol was successfully adopted, it seems that Japan's performance was not fully acknowledged by other countries.

Japan's negotiating position during the AGBM meetings was that of a middle-power country, not of a country with strong leadership. Japan was quick to respect the US position by trying to include at least some mitigation efforts from the developing countries. After adoption of the Kyoto Protocol, Japan's main focus was on the use of LULUCF and the Kyoto Mechanisms, which affected Japan's ratification of the Kyoto Protocol. The United States' withdrawal from the Kyoto Protocol in 2001, as well as internal debate within Japan over its own withdrawal from the Kyoto Protocol, shows the great consideration Japan gave to the position of the United States. At the same time, US policies towards the UN system became blurred as it took unilateral actions against terrorism, and Japanese government officials started to question their motivation to win a permanent seat on the UNSC.

Taken together, these events suggest that Japan had insufficient capacity during this time period to take a strategic lead in multilateral affairs. Political leaders were mostly inward looking, and ministries – particularly MITI and MOFA – considered the United States–Japan alliance as the core of Japanese foreign policy. Japan's position on emission reduction targets might have been quite different if the US were a country with pro-climate position had been more supportive of climate change mitigation.

Even after COP3, the Japanese government, unlike the EU, waited to see if the 6% emission reduction target would become an international commitment; the EU immediately took actions independent of the Kyoto Protocol's legal status at the international level. The position of the Japanese government was quite simple: it would act only if the Kyoto Protocol entered into force, and it would do nothing if the Kyoto Protocol was abolished at the international level. Unfortunately, there was insufficient domestic political pressure in Japan to promote climate change policies, even if the Kyoto Protocol failed to enter into force.

Nevertheless, hosting COP3 had a tremendous impact on Japanese society at the domestic level. Industries and businesses gained a much greater awareness of global environmental issues, and the establishment of environmental NGOs can almost be considered the birth of democracy in regard to environmental issues in Japan. Local governments also took important steps by implementing climate mitigation policies, even though the central government encountered difficulty in introducing key policies to start emissions reduction before the protocol was ratified. In addition, most ordinary people became aware of the terms "global warming" and "climate change," and research in climate science and economic modeling advanced to the point that it could be used to determine the country's emission reduction targets. The influence of climate change negotiations on Japan's various stakeholders should be thought of as the first tier of the climate change era in Japan.

Note

1 Maruyama, Kazuhiko (1990) *Issa Haiku shu* [Collection of Haiku by Issa]. Tokyo: Iwanami Shoten.

References

Cooper, Richard N. (2001) *The Kyoto Protocol: A Flawed Concept*, FEEM Working Paper No. 52.2001.

Global Warming Prevention Headquarters (2002) *Guideline for Measures to Prevent Global Warming*, revised. Tokyo: Government of Japan. Available online at: http://japan.kantei.go.jp/policy/ondanka/020319summary_e.html (30 January 2016).

Government of Japan (1998) *Guideline of Measures to Prevent Global Warming*. Tokyo: Government of Japan.

Government of Japan (2002) *Japan's Third National Communication under the United Nations Framework Convention on Climate Change*, submitted to UNFCCC Secretariat. Tokyo: Government of Japan.

Grubb, Michael, with C. Vrolijk and D. Brack (1999) *The Kyoto Protocol: A Guide and Assessment*. London: Earthscan, The Royal Institute of International Affairs.

Grubb, Michael and Farhana Yamin (2001) "Climatic Collapse at The Hague: What Happened, Why, and Where Do We Go from Here?" *International Affairs*, 77(2), 261–276.

Harasawa, Hideo and Shuzo Nishioka, eds. (1997) *Chikyuu Ondanka to Nippon* [Global warming and Japan]. Tokyo: Kokin Shoin.

Harris, Paul (2000) "Climate Change: Is the United States Sharing the Burden?" in P. Harris ed., *Climate Change & American Foreign Policy*. New York: St. Martin's Press, 29–50.

Hasegawa, Yuichi (2004) *Nippon Gaiko no Aidentiti* [Identity of Japanese Foreign Policy]. Tokyo: Nansousha.

Hook, Glenn D., Julie Gibson, Christopher W. Hughes, and Hugo Dobson (2012) *Japan's International Relations: Politics, Economics and Security*. Abingdon: Routledge.

Intergovernmental Panel on Climate Change (IPCC) (2001) *Climate Change 2001: Synthesis Report*. Geneva: IPCC.

Kameyama, Yasuko (2002) "Climate Change and Japan," *Asia-Pacific Review*, 9(1), 33–44.

Kawashima, Y. (2000) "Japan's Decision-Making about Climate Change Problems: Comparative Study of Decisions in 1990 and in 1997," *Environmental Economics and Policy Studies*, 3(1), 29–57.

Keidanren (1997) *Keidanren Voluntary Action Plan on the Environment*. Tokyo: Keidanren.

Kobayashi, Hikaru (2012) "Kankyo Seisaku wo Kakushin shizuketa Senkensei [Foresight that stimulated Reformation of Environmental Policies]," in Hashimoto Ryutaro Henshu Iinkai ed., *Seijika Hashimoto Ryutaro* [Hashimoto Ryutaro, a politician]. Tokyo: Bungei Shunju, 350–357.

Matsumura, Hiroshi (2000) *Japan and the Kyoto Protocol: Conditions for Ratification*, Energy and Environment Programme, The Royal Institute of International Affairs.

Ministry of the Environment (MOE) (2001) *Statement on Climate Change by Yoriko Kawaguchi, Minister of the Environment*. Available online at: https://www.env.go.jp/en/earth/cc/010409.html (accessed 30 July 2015).

Ministry of Foreign Affairs (MOFA) (1995) *Statement by Prime Minister Tomiich Murayama "On the Occasion of the 50th Anniversary of the War's End"*. Available online at: http://www.mofa.go.jp/announce/press/pm/murayama/9508.html (accessed 31 July 2015).

Oberthür, Sebastian and Hermann Ott (1999) *The Kyoto Protocol: International Climate Policy for the 21st Century*. Berlin: Springer.

Ohki, Hiroshi (2007) *Kireina Chikyu wa Nihon kara: Kankyo Gaiko to Kokusai Kaigi* [Clean Earth Starts from Japan: Environmental Diplomacy and International Conferences]. Tokyo: Genshobo.

Prime Minister's Office (PMO) (1997) *Chikyu Ondanka Mondai ni Kansuru Yoron Chosa* [Public Opinion Poll concerning Global Warming], June 1997. Available online at: http://survey.gov-online.go.jp/h09/ondan.html (accessed 31 July 2015).

Schreurs, Miranda A. (1997) "Domestic Institutions and International Environmental Agendas in Japan and Germany," in M. Schreurs and Elizabeth Economy eds., *The Internationalization of Environmental Protection*. Cambridge: Cambridge University Press, 131–161.

Schreurs, Miranda A. (2002) *Environmental Politics in Japan, Germany, and the United States*. Cambridge: Cambridge University Press.

Takahashi, Yasuo (2006) "Dai 3 sho: Beikoku no ridatsu to rekishiteki goi [Chapter 3: Withdrawal by the United States and Historical Agreement]," in Hironori Hamanaka ed.,

Kyoto giteisho wo meguru kokusai kosho [International Negotiation surrounding Kyoto Protocol]. Tokyo: Keio Gijuku Daigaku Shuppankai, 59–112.

Takamura, Yukari (2012) "Japan," in Richard Lord QC, Silke Goldberg, Lavanya Rajamani, and Jutta Burnnée eds., *Climate Change Liability: Transnational Law and Practice*. Cambridge: Cambridge University Press, 206–242.

Takeuchi, Keiji (1998) *Chikyu Ondanka no Seijigaku* [The Politics of Global Warming]. Tokyo: Asahi Sensho.Tanabe, Toshiaki (1999) *Chikyu Ondanka to Kankyo Gaiko* [Global Warming and Environmental Diplomacy]. Tokyo: Jijitsushinsha.

Tiberghien, Yves and Miranda A. Schreurs (2007) "High Noon in Japan: Embedded Symbolism and Post-2001 Kyoto Protocol Politics," *Global Environmental Politics*, 7(4), 70–91.

Torvanger, Asbjørn and Odd Godal (1999) *A Survey of Differentiation Methods for National Greenhouse Gas Reduction Targets*, TemaNord 2000:551, Nordic Council of Ministers, Copenhagen 2000.

UNFCCC (1995) Decision1/CP.1, The Berlin Mandate: Review of the Adequacy of Article 4, Paragraph 2(a) and (b), of the Convention, Including Proposals Related to a Protocol and Decisions on Follow-Up, FCCC/CP/1995/7/Add.1.

UNFCCC (1996a) The Geneva Ministerial Declaration, FCCC/CP/1996/15/Add.1.

UNFCCC (1996b) Implementation of the Berlin Mandate: Proposals from Parties, Note by the Secretariat, FCCC/AGBM/1996/Misc.2.Add.4.

UNFCCC (1997) Implementation of the Berlin Mandate: Proposals from Parties, Note by the Secretariat, FCCC/AGBM/1997/Misc.1/Add.6.

UNFCCC (1998) Buenos Aires Plan of Action, Decision1/CP.4, FCCC/CP/1998/16/Add.1.

UNFCCC (2001a) Decision 5/CP.6 The Bonn Agreement on the Implementation of the Buenos Aires Plan of Action, FCCC/CP/2001/5.

UNFCCC (2001b) The Marrakesh Accords, FCCC/CP/2001/13/Add.2.

Watanabe, Rie (2011) *Climate Policy Changes in Germany and Japan*. Abingdon: Routledge.

4 Struggling to find the "post-Kyoto" regime, 2002–2010

Overview

Tooyama ni hi-no atari taru kareno kana
(A hill far away / where sunlight shines / myself surrounded by withered field)

Kyoshi Takahama

On a cold day in midwinter in the early twentieth century, Kyoshi Takahama saw a hilltop far away glowing brightly in the sunshine and felt warm and cozy even though the surrounding field was cold. Even though the term "withered field" (*kareno*) connotes winter, the poem still evokes the warm atmosphere of the landscape.

The Kyoto Protocol entered into force in 2005, at which time Japan's emission reduction target also became legally binding. Meanwhile, Japan had started to shift its attention to the post-2012 regime, in which all major greenhouse gas (GHG) emitters were expected to participate. Non-UN processes, such as the Group of Eight (G8) process, were being utilized to stimulate political motivation to take action on climate change. Internally, the Japanese government initiated a series of decision-making processes to set an emission reduction target for 2020. Despite Japanese stakeholders' consensus that Japan should continue to insist on participation by all major emitters, the GHG emission reduction targets for 2020 for Japan, suggested by various stakeholders, were quite different from one another. Some asserted that Japan would not be able to reduce GHG emission substantially at low costs, while some others asserted that Japan would need to show enough contribution in emission reduction if Japan wanted to convince other countries to participate in a multilateral agreement.

The abolishment of the Kyoto Protocol system in which emission reduction targets were set only for Annex I countries became a kind of common goal for a majority of Japanese stakeholders. Thus, while Japan was making vigorous efforts to achieve the 6% reduction target of the Kyoto Protocol and thus fulfill its international commitment, Japanese negotiators were heading into negotiations on a new international instrument that would be able to succeed the Kyoto Protocol. A similar idea was shared by other Annex I countries that were parties to the Kyoto Protocol. The EU hesitated to abolish the Kyoto Protocol completely because it anticipated objections from the developing countries, but many other Annex I Kyoto Parties thought that any international agreement to reduce emissions beyond 2012 would need to secure the participation of all major emitting countries. The negotiation process aimed at an agreement on a new outcome that could replace, or be substituted for, the Kyoto Protocol at the 15th Conference of the Parties (COP15) in 2009.

Given that it was Japan's hope to achieve a new international agreement to replace the Kyoto Protocol and get every country to participate, in this chapter I explore whether Japan was able to find that warm hilltop filled with sunshine that could attract all of the emitting countries.

Debates over the "post-Kyoto" regime in Japan

Although Japan and many other countries ratified the Kyoto Protocol after the adoption of the Marrakesh Accord, it was still uncertain whether or not the Kyoto Protocol would ever enter into force. Article 25 of the Kyoto Protocol stated that "this Protocol shall enter into force on the ninetieth day after the date on which not less than 55 Parties to the Convention, incorporating Parties included in Annex I which accounted in total for at least 55% of the total CO_2 emission from 1990 of the Parties included in Annex I, have deposited their instruments of ratification, acceptance approval or accession." These two conditions, one on number of parties, and the other on percentage of total CO_2 emissions, were carefully chosen so that major Annex I countries needed to approve the protocol, and, at the same time, no single country – not even the United States – would have veto power against its enforcement. For this condition to be fulfilled, almost all Annex I countries had to ratify, because it was already obvious that the United States would not participate.

COP8 was held in New Delhi, India, in October–November 2002. This meeting immediately followed the significant achievements of the previous meeting in Marrakesh, so there was less enthusiasm for the meeting. Countries were obliged to start discussing what had been negotiated at COP1 in Berlin some time ago; in other words, they started discussing the specific things that needed to be done by various countries in the next round of negotiations. The host country, India, prioritized an agenda related to the interests of the developing countries. The initial version of its proposed document for the "Delhi Declaration" did not refer to the Kyoto Protocol; instead, it emphasized agenda items such as sustainable development, common but differentiated responsibilities, adaptation, technology transfer, and the commitments of Annex I parties under the United Nations Framework Convention on Climate Change (UNFCCC). Japan, together with other Annex I countries that had ratified the Kyoto Protocol, asserted the importance of the Kyoto Protocol's early entry into force, but this remained a minor issue at COP8. The finally agreed-upon text for the "Delhi Ministerial Declaration on Climate Change and Sustainable Development" called for the parties that had not yet ratified the Kyoto Protocol to do so in a timely manner. It retained all of those elements related to the developing countries' interests that had been suggested by the Indian government (UNFCCC 2002).

The Japanese government did not put much effort into helping get the Kyoto Protocol entered into force. While the EU negotiated with Russia to ratify the Kyoto Protocol, some domestic protagonists in Japan hoped that Japan's 6% emission reduction commitment would be nullified. As they waited for the Kyoto Protocol to enter into force, Japanese ministries began discussions on what should be done after 2012 – that is, at the end of the first commitment period of the Kyoto Protocol. The Ministry of Economy, Trade, and Industry (METI) and Ministry of the Environment (MOE) were particularly interested in this debate. The two ministries used their respective committees to organize meetings and draft reports on this subject.

In July 2003, METI's Environmental Committee under the Industrial Structure Council published an interim report of their debate on the future framework for climate change negotiations (Industrial Structure Council 2003). The report asserted four fundamental bases on which a future international framework on climate change should stand:

• the need for technological innovation
• a diversified agenda in each nation, region, and sector
• the need for an awareness of the tremendous global cost of mitigation policies
• a better understanding of remaining scientific uncertainties.

The report emphasized scientific uncertainties, stating, "[T]he mechanisms and effects of climate change still have significant uncertainties." It also stated, "[P]revention of global warming would require the world to bear enormous costs for measures." Thus, the report expressed the view that a major technological breakthrough was indispensable, saying that "a future framework should take into account development and dissemination of innovative technology related to mitigation of climate change." Such technology needed to be developed and disseminated mainly on a voluntary basis, because "the actions required to prevent global warming may vary widely depending on the specific situation of each nation, region, and sector, and there were very significant variations in the costs for those actions."

In a parallel manner, the MOE also requested the Global Environment Committee, under the Central Environment Council, to publish an interim report in January 2004 (Central Environment Council 2004a). This report suggested the following seven fundamental elements as the way to move forward:

• maintain progress toward meeting the ultimate objectives of the UNFCCC
• bring the Kyoto Protocol into force and fulfill commitments
• achieve global participation
• ensure equity on the basis of the principle of common but differentiated responsibilities
• build future negotiations on existing international agreements
• formulate an international consensus-building process by national governments with the participation of various actors
• make the environment and economy mutually reinforcing.

The report referred to the Third Assessment Report of the Intergovernmental Panel on Climate Change (IPCC) and concluded that "relevant scientific work over several decades has reduced scientific uncertainty." The Kyoto Protocol was considered a significant first step toward meeting the ultimate objective of the UNFCCC, which was the stabilization of GHG concentrations in the atmosphere at a level that would prevent dangerous anthropogenic interference with the climate system. As for the next round of negotiations for the years after 2012, the report considered that "ensuring environmental integrity of the climate regime requires global participation," and therefore that "the climate regime beyond 2012 needs to achieve the participation of all countries, including the United States and developing countries." Finally, the level of participation needed to be differentiated on the basis of the principle of common but differentiated responsibilities.

Both METI's Environmental Committee and MOE's Global Environment Committee considered the participation of the United States and major developing countries to be indispensable in the next round of negotiations. This was particularly the case because noncommitment by these countries on emission mitigation was consistently noted by critics of the Kyoto Protocol. Both reports also acknowledged the roles of the economy and technology, and they encouraged incentives for countries and domestic actors to shift toward more climate-friendly actions.

Despite the commonalities, there was still a large gap between the two reports on the following issues: the perception of the degree of scientific uncertainty and how to deal with it; the process needed to move forward; the extent to which the concept of equity should be considered important; and the role of governments. These differences led the ministries to draw two different conclusions on how the UNFCCC regime should evolve to deal with the climate problem in the future. These divergent views generally reflected the two prevailing views of the Kyoto Protocol in Japan at that time.

As long as the two ministries primarily in charge of climate change policy had such divergent views, particularly because the Kyoto Protocol had yet to enter into force, the MOE was unable to implement its ambitious emission reduction policies to curb Japan's GHG emissions. One of the policy instruments that actually was available to the MOE at the time was a voluntary emissions trading scheme (ETS). When Japan ratified the Kyoto Protocol, the MOE started pilot-phase emissions trading in 2003 and 2004. It started operating a simple ETS on a voluntary basis so that participants could gain practical experience with this type of system and the MOE could learn from those experiences. By using the lessons learned from these experiences, the MOE intended to be able to fully implement an ETS at about the same time as the review of the *Guidelines of Measures to Prevent Global Warming*, which was scheduled in 2004. When the Kyoto Protocol entered into force in 2005, the pilot phase of the ETS shifted to become the Japan Voluntary Emissions Trading Scheme (JVETS). Nearly 300 companies voluntarily participated in the scheme. This was not, however, a real cap-and-trade scheme, because the cap was set voluntarily by industry. There had been strident opposition from industry against setting an absolute emission cap, so the government could not introduce anything even remotely close to a real cap-and-trade ETS (Global Warming Prevention Headquarters 2008). As the result, few companies required use of JVETS to achieve their respective targets, while other companies made small transactions only on a trial basis. The maximum participation to the trading was 86 companies, and total transaction accounted for a maximum of 51 transaction per year (Rudolph and Schneider 2013).

Another conflict between the MOE and METI involved the introduction of environmental taxes. In 2003, then environment minister Shunichi Suzuki instructed the MOE to start an internal investigation on the introduction of an environmental tax in anticipation of drafting a bill on this subject. After coming up with a draft proposal on legislation for an environmental tax, in November 2003, the MOE asked the Environment Working Group of the Liberal Democratic Party (LDP) to discuss the bill and pass it in the following fiscal year. At the time, LDP members were reluctant to discuss a new tax and decided not to pursue this issue.

The MOE did succeed, however, in gaining ground in discussions of an existing tax. As a part of a general restructuring of government institutions, an existing account known as the "oil special account," derived from an earmarked tax of which the revenues were used to subsidize the coal industry and oil-related organizations, was reconsidered. METI was administratively in charge of the tax revenues, some of which had started to be used (since 1993) for other purposes, such as investment in renewable energy and improvements in energy efficiency. In 2003, use of the tax revenues was further extended to include aid programs aimed at reducing CO_2 emissions from fossil fuel combustion, and the MOE, together with METI, now had power over the distribution of these tax revenues. Environment Minister Suzuki is said to have negotiated to win some power in spending this tax revenue to support emissions reduction, but the trade-off was backing off on the issue of a carbon tax.

Waiting for the Kyoto Protocol to enter into force

At the international level at this time, a wide variety of proposals were being made on what type of international institution would be desirable in the post-2012 period (Bodansky 2004; Kameyama 2007, 2008). One side argued that emission reduction targets should be set for all countries – that is, some sort of continuation of the Kyoto Protocol would be desirable (Den Elzen et al. 2003; Höhne et al. 2003; Meyer 2000; Ott et al. 2004). Proposals based on this idea were known as "post-2012" or "beyond-2012," meaning that something – but not necessarily the entire protocol – had to be revisited after the protocol's first commitment period. Meanwhile, some others stressed the failure of the Kyoto Protocol and indicated that something completely different should be envisaged (Nordhaus 2001; Schelling 2002; Stewart and Wiener 2003; Victor 2004). Proposals based on this idea were known as "post-Kyoto," reflecting the need for a complete restart. Still other proposals focused on ways in which developing countries could be integrated into mitigation actions (Bosi and Ellis 2005; Bradley and Baumert 2005).

As researchers and negotiators were busy discussing what was to be agreed upon in the next round of negotiations, COP9 was held in Milan, Italy, in December 2003. The parties were still waiting for the Kyoto Protocol to enter into force. This required the participation of Russia, but Russia was determined to get the most it could out of the COP9 negotiations before it would ratify the protocol.

Most of the agenda items at COP9 were quite technical. For example, the inclusion of forest-related projects under the Clean Development Mechanism (CDM) was one issue. At the Marrakesh Accord, the parties agreed that some project activities related to afforestation and reforestation could be applicable as CDM projects. Detailed rules concerning the definitions of reforestation, leakage, baselines, and nonpermanency, as well as the social and environmental effects of projects, were discussed. Forest-related CDMs were considered relatively inexpensive compared with other projects such as those related to renewable energy, so some countries were concerned that the amount of certified emission reductions (credits) should be limited in the case of simple CDM projects. Compiling emissions inventories was also another rather technical agenda item, but it was essential, particularly so that Annex I countries could prove that they had met (or not met) their commitments under the Kyoto Protocol.

Emissions arising from international bunker fuels were not included in national inventories, but some small island states and the EU asserted that emissions from this sector were increasing, and that some rules should be set in place to allocate emissions from that sector to countries. Other countries, including Japan, were reluctant to include emissions from international bunker fuel in their own inventories, because this could substantially change each country's total emissions, depending on how the emissions were allocated.

Three new funds were established in the Marrakesh Accord, but negotiations on the details continued throughout COP9. The developing countries were anxious to see a flexible funding institution, with additional funding resources to be used mainly for adaptation and technology transfer rather than mitigation actions. Japan took the position that at least a minimal procedure such as submission of application documents was important to ensure that the financial resources would be used in the most effective manner. Japan and other developed countries hoped to start formal discussions on what should be done after the end of the first commitment period, but the developing countries were not interested in having these discussions.

The *Guideline of Measures to Prevent Global Warming* (revised in 2002) set by the Japanese government stated that the first review of the guideline be conducted in 2004. This review process was considered particularly important by pro-environment groups, because the 2002 guidelines called for the use of a "step-by-step" approach, which implicitly meant postponing any meaningful policies or measures until the 6% emission reduction target actually became legally binding under the Kyoto Protocol. Japan's GHG emissions had grown in the

Table 4.1 Japan's GHG emissions in fiscal year 2002

Sector	Target reduction in the 2002 revised guideline	Actual emissions	Gap between target and actual emissions
CO_2 emissions from energy combustion	0%	+12%	+12%
Industry	–7%	–1.7%	+5.3%
Residential	–2%	+28.8%	+30.8%
Commercial	–2%	+36.7%	+38.7%
Transportation	+17%	+20.4%	+3.4%
Other CO_2, CH_4 and N_2O	–0.5%	–1%	–0.5%
HFCs/PFCs/SF_6	+2%	–2%	–4.0%

HFCs: hydrofluorocarbons; PFCs: perfluorocarbons (PFCs); SF_6: sulfur hexafluoride

Source: MOE (2004)

1990s, and small ups and downs were observed in the years after the late 1990s. There was no evidence of a declining trend in national emissions.

Emissions in fiscal year 2002 are shown in Table 4.1. The total amount exceeded the base year level by 7.6%. This meant that emissions had to be reduced by more than 13% of the level in 1990 to meet the 6% reduction target. A breakdown of emission levels into various sectors indicated a significant increase in the residential and commercial sectors. On the basis of these data, industry groups asserted that the industrial sector was well on its way towards successfully meeting its sector target and that emissions from the residential and commercial sectors should be of more concern. They also maintained that the emission trends justified the use of their voluntary action approach.

As the Central Environment Council worked toward revising the guidelines, Russia finally adopted the Kyoto Protocol in November 2004, and it became apparent that the Kyoto Protocol would enter into force in February 2005. The final report of the Central Environment Council urged the Japanese government to take additional measures to ensure meeting the 6% reduction target. It included measures such as the use of environmental taxes, a cap-and-trade type of ETS among relatively large emitting industries and companies, and tax exemptions for low-emissions products. The committee also recommended that the MOE oblige companies to report their levels of GHG emissions to the government each year. The Industrial Structure Council under METI opposed environmental taxes and asserted that only further revisions to the *Law Concerning the Rational Use of Energy* were needed to meet the 6% reduction target. The Industrial Council was also reluctant to implement a reporting requirement, because the same *Law Concerning the Rational Use of Energy* already had instructed companies to submit data on energy consumption; the council argued that the two similar obligations represented a duplication of effort.

Finally, a compromise position was reached. The environmental tax was turned down, but reporting requirements for GHG emissions were included. The review of the guidelines resulted in the Kyoto Protocol Achievement Plan, which was agreed upon within the Japanese government in April 2005. Although this review was considered an opportunity for the pro-climate policy group to introduce an additional policy instrument, it faced strong opposition from industry groups, resulting in further postponement of the introduction of any new measures or instruments.

While the Japanese government was having internal discussions on how to respond to the Kyoto Protocol's entry into force, COP10 was held in Buenos Aires in December 2004. Countries applauded Russia for its ratification of the Kyoto Protocol. Now that the parties were sure that the next COP would be held in conjunction with the Meeting of the Parties to the Kyoto Protocol, negotiators had a more positive outlook in discussing the future of the climate regime.

One of major agenda items at COP10 was to start the debate on the future regime – that is, the regime after the first commitment period of the Kyoto Protocol. The EU, Japan, and other Annex I parties to the Kyoto Protocol, as well as the host country, Argentina, wanted to start the discussion by holding a seminar, hoping such dialogue would lead to starting a formal negotiation. The United States and many developing countries, however, were reluctant to begin such talks, preferring to hold a single-shot side event rather than a seminar. Eventually, the parties agreed that a seminar would be held in conjunction with the next subsidiary body meeting in June 2005 but that the result would not be reported to COP11.

What became even more important in the COP10 meeting was the agenda on adaptation. Although developing countries had sought more meaningful progress in the area of adaptation, developed countries had not been fully cooperative. COP10 adopted the "Buenos Aires Program of Work on Adaptation and Response Measures," which requested the Subsidiary Body for Scientific and Technological Advice to develop a structured five-year program of work on the scientific, technical, and socioeconomic aspect of impacts, vulnerability, and adaptation to climate change (UNFCCC 2004).

During COP10, METI held a side event separate from the formal negotiation meetings. Although the title of the event mentioned developing countries' energy-related issues, it was actually a presentation of the interim report of METI's Committee on the Future Framework under the Industrial Structure Council. Although this report represented only the position of one ministry, it was presented as if the report represented the Japanese government as a whole, thereby sending a misleading message to the audience.

The Kyoto Protocol enters into force

Russia finally ratified the Kyoto Protocol in November 2004, and this led to the Kyoto Protocol's entry into force in February 2005. Consequently, the Japanese cabinet adopted the *Kyoto Protocol Target Achievement Plan* in April 2005. This plan, together with the revised guidelines, provided a list of planned measures for climate change mitigation at the national level. Most measures that were introduced under the law were voluntary, including the Keidanren Voluntary Action Plan, which was the core part of Japanese mitigation measures in the industrial sector (Keidanren 2008). In this action plan, each industry association voluntarily pledged a target, which was to be reviewed by the government on an annual basis. The measure was voluntary in two respects. First, each industry association and each company belonging to the association voluntarily participated in the plan. Second, the form of the target (absolute or relative) and the level of the target were determined by the participating companies themselves.

Most of the mandatory measures in Japanese climate policy were the ones taken to improve energy efficiency under the *Law Concerning the Rational Use of Energy*, enacted in 1979 and later amended. Recommendations to introduce more stringent mandatory climate policies were raised at the time the plan was drafted and each time it was periodically reviewed, but there was never enough support from the industrial sector to revise the law to include more stringent or mandatory climate policies.

The MOE started implementing measures to reduce GHG emission. One of the actions taken was "Cool Biz," a campaign that encouraged people to wear light clothes during the hot summers. Cool Biz did not require men to wear ties or jackets (a dress code that was typical for Japanese businessmen, even during summer). The purpose of this campaign was to get people to set the thermostat at 28 °C when using air-conditioners. The environment minister at the time was Yuriko Koike, who had been a journalist before she was elected to the Diet. Her past experience in mass media was advantageous in creating and marketing Cool Biz – for example, by requesting top business leaders to appear as models in a series of Cool Biz fashion shows (Sampei and Aoyagi-Usui 2009). Warm Biz was similarly successful in the winter.

In the first half of 2005, some major economies took significant actions even before the Kyoto Protocol's entrance into force. For example, the EU implemented an ETS in January 2005. This was a decisive moment for the EU because, once the government had established a carbon market, it was important to continue sending messages to industries and business communities that the market would be reliable and continue to exist far into the future (Skjærseth and Wettestad 2008). For the EU's regional ETS to continue, it would be beneficial for the EU to have the Kyoto Protocol extend into the post-2012 era and to set emission reduction targets for other Annex I parties under the Kyoto Protocol.

The United States also saw various actions taken internally. First, in the Congress, various climate-related bills were proposed. Among those were the Climate Stewardship Acts of Senator John McCain and Senator Joseph Lieberman in 2003 and 2005, respectively, which aimed to set emission caps on GHGs or CO_2 and introduce a cap-and-trade ETS in the United States. None of the bills won a majority within the Congress, but they were viewed as at least an attempt to meet the spirit of the Kyoto Protocol, to which the United States was not a party.

More progress was made at the state level. California, which has long been a pioneer in addressing environmental problems, started to address CO_2 emissions reduction in the early 2000s. In June 2005, Governor Arnold Schwarzenegger announced an executive order setting a state-level GHG emission reduction target of 11% by 2010, 25% by 2020, and 80% by 2050 relative to the 2005 level. The following year, California set up the Global Warming Solutions Act. The Chicago Climate Exchange (CCX) was established in 2003 to voluntarily establish a comprehensive cap-and-trade program with an offset component. More than 400 companies, municipalities, and educational institutions participated in the trading program. Trading under the CCX lasted until 2010, at which point the CCX launched an offset registry program.

At the federal level under the Bush administration, however, the United States was not very enthusiastic about joining another legally binding treaty under the UNFCCC that could hamper growth in the US economy. It preferred an alternative regime outside the UN arena. In June 2005, the United States headed the establishment of the Asia-Pacific Partnership on Clean Development and Climate (APP), involving Australia, Canada, Japan, China South Korea, and India. The aim of this organization was to cooperate in promoting the deployment of clean energy technologies to achieve the goal of energy security and climate protection. Japan was the only country among the APP members that had committed to an emission reduction target during the first commitment period of the Kyoto Protocol. By joining such an organization outside the UN system, Japan hoped to be able to control sectoral emissions from large GHG emitters that were not controlled by the Kyoto regime. As shown in Table 4.2, most of the CO_2 emissions from the most energy-intensive industries originated in countries that did not have emission reduction commitments under the Kyoto Protocol. Japan was eager to set some kind of control over production in these countries, and the APP was perceived as an excellent tool to regulate these types of emissions.

Some viewed the establishment of the APP and other local initiatives as a "fragmentation" of the climate regime (van Asselt 2014). This contributed to worries that the UNFCCC was no longer as influential as it was in the twentieth century. The UNFCCC process, however, was still widely considered to be the central pillar of the climate regime at the multilateral level.

The first COP serving as the Meeting of the Parties to the Kyoto Protocol (CMP1) was held in Montreal, Canada, in December 2005, in conjunction with

Table 4.2 Top five emitters in the steel and cement sectors (2000)

	Steel	$MtCO_2$	Cement	$MtCO_2$
1	China	290	China	500
2	Russia	91	USA	104
3	Japan	88	India	78
4	USA	75	Japan	70
5	India	59	South Korea	42

Source: Schmidt et al. (2006)

COP11. Even though this was a historic moment in which member countries celebrated the Kyoto Protocol's entrance into force, countries were also extremely interested in discussing how to start a new negotiating process. Japan and many Annex I Kyoto parties insisted on starting a new process to negotiate a new institutional framework for the years after the first commitment period – that is, after 2012. There were two relevant articles in the Kyoto Protocol relating to this. Article 3.9 stated, "commitments for subsequent periods for Parties included in Annex I shall be established in amendments to Annex B to this Protocol, which shall be adopted in accordance with the provision of Article 21.7. The CMP shall initiate the consideration of such commitments at least seven years before the end of the first commitment period referred to in paragraph 1 above."

According to this schedule, a negotiation to determine Annex I countries' emission reduction targets for the second commitment period had to begin at COP11. Many Annex I countries that had ratified the Kyoto Protocol, including Japan, did not wish to start such negotiations without other countries – especially the United States and China – making similar commitments regardless of their participation in the Kyoto Protocol.

Article 9 was also relevant, because it addressed the review of the protocol, stating, "the first review shall take place at CMP2." The Annex I countries recognized that this review meant reconsideration of countries that were to commit on emission reduction targets and that this agenda could be addressed only at the next CMP, not at CMP1.

Other countries, including the United States and many non–Annex I countries, whose GHG emissions were not bound by the Kyoto Protocol, were reluctant to start new formal negotiations to achieve a legally binding document that included legally binding emission reduction commitments for themselves. After a long debate, the COP adopted a decision to start a two-year dialog, not a formal negotiation, to work out what countries would do for the years beyond 2012. Four workshops were planned during the dialog, but they would not lead to a formal negotiation (UNFCCC 2005a).

A new negotiation stage to determine the second commitment period of the Kyoto Protocol was started under the CMP. This process was called the "Ad hoc Working Group on Further Commitments for Annex I Parties Under the Kyoto Protocol" (AWG-KP) (UNFCCC 2005b). The decision stated that the parties "[a]gree that the group shall aim to complete its work and have its results adopted by the CMP as early as possible and in time to ensure that there is no gap between the first and second commitment periods." Although there was no indication of the year in which this process would be concluded, negotiators recognized that the process needed to be concluded by 2009 or 2010, at the latest, for countries to be able to ratify the amendments and still have no gap between the first and second commitment periods. Confirmation of the deadline was crucial for the EU, because it wanted to send a message to its regional carbon markets that the market was going to continue in existence even after 2012 to ensure its stability (Azar 2005). Meanwhile, Japan and some other industrialized countries were in no hurry to make much progress under the Kyoto Protocol and the COP unless all countries were going to be included. In the end, these agreements were put into a package called the "Montreal Action Plan."

The year 2006 passed without much upheaval on the climate change issue, both internationally and domestically for Japan. The country's GHG emissions were not within reach of the 6% emission reduction target, and the beginning of the commitment period was only two years away. Industry groups continued to criticize the Kyoto Protocol as ineffective and to contend that a 6% reduction was unachievable. They asserted that Japan should abolish the Kyoto Protocol after 2012 and start a new regime beginning in 2013 – if not withdraw from the protocol altogether.

COP12/CMP2 was held in Nairobi, Kenya, in November 2006. Japan's main interest was in Article 9 of the Kyoto Protocol, which stated, the "CMP shall periodically review this Protocol. . . . Based on these reviews, the CMP shall take appropriate action," and "[t]he first review shall take place at CMP2." Japan's view was that this article represented the only official avenue available to start negotiations on emission mitigation commitments for non–Annex I countries. Japan's position, however, was strongly opposed by the non–Annex I group, which held the position that this type of review should not be part of any new negotiating phase. The developing nations thought that the participation of non–Annex I countries was within the mandate of the dialog process set up under the COP and that setting emission reduction targets for developed countries was the mission of the Kyoto Protocol. After all, they agreed that the first review to conclude without starting a new negotiation, and that the second review to be conducted in 2008.

Kenya was the host country, and the conference agenda was tilted toward the interests of the developing country group. Negotiators discussed the Adaptation Fund under the Kyoto Protocol and adaptation plans in the developing countries. The COP decision, the "Five-year Program of Work of the Subsidiary Body for Scientific and Technological Advice on Impacts, Vulnerability and Adaptation to Climate Change," also known as the "Nairobi Work Program," was adopted for developing countries to start preparations for the adverse impacts of climate change. In terms of the CDM, the African Group had been concerned that most CDM projects were being hosted by emerging economies such as China and India. The group had hoped that CDM projects would be more equally distributed. The CMP2 considered this concern and agreed on the "Nairobi Framework," which aimed at capacity building in developing countries to increase the number of CDM projects.

Another interesting agenda item – proposed by Brazil at COP11 – was on reducing emissions from deforestation in developing countries. Countries started discussing the proposal by holding workshops and inviting submissions from countries and nongovernmental organizations (NGOs). This agenda eventually developed into Reducing Emissions from Deforestation and Degradation in Developing countries (REDD) at COP13. All of these agendas were important for the climate regime to move forward at the international level, but none of them directly affected Japan's internal affairs; therefore, they attracted little attention from Japanese political leaders.

Emergence of the G8 Process (2007 and 2008)

The year 2007 started with good news for climate mitigation supporters. First, the IPCC won the Nobel Peace Prize and was lauded as an organization that was working globally to move toward more ambitious emission mitigations (IPCC 2007). Sharing the award was former US vice president Al Gore, who was praised for his work on enlightening people on the importance of climate change. In addition, the United Nations General Assembly session on climate change was held in the fall of 2007. Another positive indicator was that China, which had GHG emissions almost equal to those of the United States, had become more active in responding to climate change. In general, the high-level politicians and bureaucrats seemed to be becoming more aware of the role China must play in pushing climate change negotiations at international level.

Climate change had been a central topic of the G8 since the Gleneagles Summit in 2005. Japan's Ministry of Foreign Affairs (MOFA), which was to play a central role in the Toyako Summit to be held in 2008, was thus deeply involved in the discussions. Similarly, Japan's prime minister, who would be hosting the summit, was interested in demonstrating his leadership on the climate change issue (MOFA 2008). As a result of these political changes in the international arena, Japan's politicians were pressured to start giving greater consideration to the global climate change problem. Prime Minister Shinzo Abe had held office since September 2006, but he had rarely expressed opinions on climate change and was not known for being enthusiastic about environmental protection. In his address on administrative policy, given on 16 January 2007, he did mention a plan aimed at achieving the goals of the Kyoto Protocol and offering support to developing countries, but only after he had discussed other internal issues, including the creation of a sound and safe society. Just before giving this speech, Abe had visited European countries to seek their understanding on the abduction of Japanese nationals by North Korea. Although his intention was to talk only about North Korea, the European Commission and the individual countries he visited all were more interested in explaining the EU's proposed package of climate and energy policies. Abe eventually became aware of the importance of the issue.

Abe often used the key phrase "Beautiful Country" to express various issues, including economic and social ones. Government officials suggested the similarly worded "Beautiful Planet 50" as the title of a speech he gave in May 2007 (Prime Minister's Office 2007). The proposal outlined in this speech consisted of three key components:

(1) long-term strategies for halving global GHG emissions by 2050
(2) medium-term approaches on the framework beginning in 2012
(3) policies for fulfilling the 6% reduction commitment of the Kyoto Protocol.

The long-term objective did not clearly define the base year for the 2050 emission reduction proposal, but it still attracted the attention of political leaders from around the globe. At the G8 Summit, held in Heiligendamm, Germany, in June 2007, Prime Minister Merkel of Germany and leaders of other European nations requested that the other group members commit to halving global emissions by 2050. President Bush of the United States strongly resisted, so the summit countries announced only that they would "aim to" at least halve global CO_2 emissions by 2050, rather than making any firm commitments to the goal.

Abe most likely wanted to remain in office until the Toyako Summit in 2008, but he resigned soon after the Heiligendamm Summit. His successor, Yasuo Fukuda, had little personal interest in environmental problems but also had no reason to strongly oppose environmental measures either. Fukuda made slow but steady strategic arrangements to prepare himself for the Toyako Summit. While ministers of G8 countries met in Heiligendamm, the United States continued to seek a way to discuss climate change in forums other than those under the UN system. It hosted the "Major Economies Meeting on Energy Security and Climate" (MEM) in September 2007 and invited about 20 major countries and regions, including some non–Annex I countries. The group also met several times in 2008; this did contribute to the informal dialog on climate change among countries, but it was not influential enough to replace the UNFCCC. The MEM was later renamed the Major Economies Forum by the Obama Administration.

COP13 took place in Bali, Indonesia, in December 2007. The central debate of the negotiations at the Bali meeting was how to start a new negotiation process to reach an agreement on the future international climate regime (i.e., beyond 2012). The Japanese government's position consisted of two main points (Government of Japan 2007). The first was that the expected outcome should include emission mitigation commitments not only for the developed countries but also for developing countries, at the least for the major emerging economies such as China and India.

The second point was that commitments of the parties should include use of the "sectoral approach." At this time, Japan had not consolidated its position on whether or not to accept a quantitative emission reduction target similar to that contained in the Kyoto Protocol, largely because of the strong objections from some key industries against absolute emission targets. These industries continued to view the 6% emission reduction target as unfair, because their competitors in China, India, and the United States were not facing the same emission reduction requirements. They perceived themselves to be among the most energy-efficient companies in the world, and they did not accept the rationale behind the exclusion of these other countries. They still preferred agreeing to sector-wide, nonlegally binding, energy-efficiency targets. Because their competitors included current non–Annex I countries, they wanted to set sectoral targets in both Annex I and non–Annex I countries.

An agreement reached at COP13, the "Bali Action Plan," included a paragraph that called for a discussion on "nationally appropriate mitigation actions by developing country Parties" (UNFCCC 2007). The first point of Japan's position – to involve developing countries – was thus reflected in this paragraph. Its second point, about the use of a sectoral approach, was also reflected in the agreed-upon text, but not in a way that Japan had expected. A paragraph on the sectoral approach stated that the upcoming negotiation would address "cooperative sectoral approaches and sector-specific actions, in order to enhance implementation of Article 4, paragraph 1(c) of the Convention." This wording implied that the sectoral approach under the Bali Action Plan was intended mainly to motivate technology transfer in developing countries and not as part of their emission reduction commitments related to mitigation actions.

The most significant point of the Bali Action Plan was the initiation of a new round of negotiations to achieve an agreement on what was to be done between 2012 and 2020. The paragraph stated that the COP "decides to launch a comprehensive process to enable the full, effective and sustained implementation of the Convention through long-term cooperative action, now, up to and beyond 2012, in order to reach an agreed outcome and adopt a decision at COP15." The agreed-upon outcome was to address a shared vision for long-term cooperative action and enhanced actions on national mitigation, adaptation, technology development and transfer, and provision of financial resources. There was no agreement reached as to whether the "agreed outcome" should be expressed as a new protocol, by way of an amendment of the UNFCCC or Kyoto Protocol, or through a series of COP decisions. Even though the issue of the legal nature of the agreement was not discussed at the time, it has become an important factor in the complex negotiations (Clémençon 2008). The process that was launched by the Bali Action Plan was called the Ad Hoc Working Group on Long-Term Cooperative Action Under the Convention (AWG-LCA).

Determination of emission reduction targets for 2020: the Toyako G8 Summit

Early in 2008, interest in climate change negotiations and the Toyako G8 Summit began to increase across various elements of Japan's government. Soon after the Bali meeting, the government began discussions to consolidate Japan's position on whether or not to accept an absolute emissions reduction target at the national level. This closed meeting involved ministers from MOE, METI, MOFA, and the Ministry of Finance (MOF), so the meeting was unofficially dubbed the "four-ministers' meeting on climate policies." Most of the ministers were basically supportive of setting an emission reduction target at the national level, but the METI minister was strongly opposed to choosing this option, because he was sympathetic with the view of industry.

The view of various Japanese industrial organizations was elaborated most clearly by discussion papers published by the 21st Century Public Policy Institute, an organization with abundant funding by members of Keidanren, Japan's largest group representing industry and business (Sawa 2008). The concept of the sectoral approach had several definitions that shifted over time and among Japanese stakeholders. At the time of the negotiations up to the Bali Action Plan and even afterward, Japanese government officials and business stakeholders believed that a sectoral approach meant setting either absolute emission targets or energy efficiency targets voluntarily at the sectoral level only. Here, a "sector" is considered to be a type of industry, such as the electricity-generation or iron and steel sectors. Keidanren stressed that the sectoral approach, as they defined it, had advantages over the existing Kyoto framework because it would encourage the involvement of a wider range of countries, including non-Kyoto countries and emerging economies. In addition, they thought it would resolve a variety of issues concerning international competitiveness (Keidanren 2008). As long as key industries' emissions were controlled across borders, they argued, national reduction targets would be less important.

Then final decision was made by then prime minister Fukuda himself. In a speech he made at the annual World Economic Meeting held in Davos, Switzerland, in January 2008, he presented a proposal called the "Cool Earth Promotion Program," which basically accepted a quantitative emission reduction target at the national level (Prime Minister's Office 2008a). He noted that the national target for each country should be based on calculations of the amount of emission reduction potentials in each country. He said, "Japan will, along with other major emitters, set a quantified national target for the GHG emissions reductions to be realized from now on. In setting this target, I propose that the equity of reduction obligations be ensured. The target could be set based on a bottom-up approach by compiling, on a sectoral basis, energy efficiency as a scientific and transparent measurement and tallying up the reduction volume that would be achieved based on the technology to be in use in subsequent years." This methodology has since become the new definition of the "sectoral approach" inside the Japanese government.

This proposal also was made at the negotiation meetings held in Bangkok and the Chiba G20 Environment and Energy Ministerial Meeting both in March 2008. Developing countries strongly objected to the proposal, because it suggested setting emission reduction targets for all major emitting countries, including large emerging economies such as China and India.

During this period, the EU continued in its role as the leader of the climate change dialog at the international level. Already, in March 2007, the leaders of the EU had agreed at the European Council on setting ambitious emission reduction targets for 2020. In January 2008, the European Commission published a communication called "20 20 by 2020 Europe's Climate Change Opportunity" (European Commission 2008). The document called for a GHG emission reduction of at least 20% by 2020, rising to 30% if there was an international agreement committing other developed countries to "comparable emission reductions and economically more advanced developing countries to contributing adequately according to their responsibilities and respective capabilities." It also called for a 20% share of renewable energy in total EU energy consumption by 2020, in addition to the "20% improvement in the EU's energy efficiency" that was agreed upon in the EU Climate and Energy Package to be enacted in 2009.

Japan, meanwhile, continued to make progress internally in preparing for the Toyako G8 Summit. A "Council on the Global Warming Issue" was established under the Prime Minister's Office in February 2008. The council consisted of 12 members from academia, the business community, or other related organizations, and it met once a month until the G8 Summit in July 2008. The council discussed various aspects of the issue, including long-term targets, a vision for a low-carbon society, and technology transfer, among other issues. During the Abe administration a year earlier, Japan had already proposed halving global emissions by 2050. The next step was to investigate a long-term emission target for Japan. In early June, Fukuda announced the "Fukuda Vision," which included Japan's emission target for 2050. He said that Japan was ready to aim at reducing its emissions by 60%–80% of the current level by 2050 and to develop a low-carbon society. The vision called for a stronger position on climate policy and included a domestic ETS (Prime Minister's Office 2008b); the trading scheme to be introduced on a trial basis in October 2008.

The Council on the Global Warming Issue also published its own "Proposal of the Council on the Global Warming Issue, in Pursuit of Japan as a Low-Carbon Society" (Council on the Global Warming 2008). The council basically supported the Fukuda vision by highlighting key points, such as a 60%–80% emission reduction from the current level by 2050, technology innovation, use of low-carbon energy resources, and carbon pricing.

The Toyako Summit was held in July 2008; global warming was discussed. Prime Minister Fukuda and MOFA were actively encouraged to participate in the discussion and in making decisions on measures to tackle global warming. The meeting highlighted the need for an agreement on the 2050 global target, but the United States was not willing to accept a clearly defined emission-level reduction as a target to be achieved by the G8 member countries. To reflect all countries' concerns, the final chair's summary wrote, "we seek to share with all Parties to the UNFCCC the vision of, and together with them to consider and adopt in the UNFCCC negotiations, the goal of achieving at least a 50% reduction of global emissions by 2050, recognizing that this global challenge can only be met by a global response, in particular, by the contributions from all major economies, consistent with the principle of common but differentiated responsibilities and respective capabilities" (MOFA 2008).

Determination of emission reduction targets for 2020: after the Toyako Summit

Soon after the Toyako G8 Summit, Fukuda suddenly resigned his position as Japanese prime minister. He was succeeded by Mr. Taro Aso. Prime Minister Aso generally followed the future climate policies that had been determined under the Abe and Fukuda administrations. Aso also maintained the Council on the Global Warming Issue, which met again in October 2008. Two significant items were discussed at this meeting. The first concerned the introduction of a domestic ETS. The Japanese government had not yet been able to introduce an ETS at the domestic level, primarily because of the strong objections of various industries. Meanwhile, an ETS had already been introduced at the domestic level in the EU and had been at least partially introduced in the United States and some other developed countries. Two distinct committees were established to discuss the scheme, one under MOE and the other under METI. Thus, the council played a role in paving the way to start a trading scheme in Japan.

Japan started a pilot phase for an ETS in October 2008, as had been decided by Prime Minster Fukuda. The MOE planned a transition period before the ETS would become mandatory in 2013. This was a voluntary ETS, in an attempt to bring the companies currently operating under the Keidanren's Voluntary Action Plan (1997) into the ETS. In the plan, industry associations and individual companies could decide to adopt an absolute or relative emissions target and could independently determine the level of the target. Verification of the participants' emissions was not required unless a company wished to sell excess allowances. Given these features, it was highly unlikely that a significant number of allowances would be traded or that a positive carbon price would evolve. The transition to a mandatory cap-and-trade scheme faced strong opposition by Japanese industry from the beginning of the pilot phase.

Another important item approved by the council was the establishment of the "Committee on the Mid-term Emission Target" under the council. Eleven years before this, when Japan internally discussed the emission target to be incorporated in the Kyoto Protocol in 1997, joint meetings of the Central Environment Council and Industry Structural Council were held several times, but the substantive internal negotiations were conducted between ministries (especially the MOE, METI, and MOFA) behind closed doors. The joint meetings of the two councils did not affect the final agreement among the three ministries at the time. Unlike in the previous process, this new committee would be the decision-making body, and it was established inside the Prime Minister's Office, separately from the MOE and METI. This newly established committee was tasked with using several economic models to develop scenarios that reflected Japanese people's concerns, such as by how much Japan could reduce GHG emissions at specified costs and how a given level of effort could be compared with those of the EU or the United States. The developed scenarios were used as references for the Prime Minister to make the final decision on Japan's midterm emission reduction target. An agreement made at COP14, held in December 2008 in Poznan, Poland, invited parties to present ideas on their relevant emission reduction targets by the next AWG-LCA meeting, to be held in March 2009, so the Committee on the Mid-term Emission Target needed to move relatively quickly to complete its task.

The committee met seven times, between November 2008 and April 2009. It consisted of eight high-level experts in the fields of energy, the environment, and economics (MOE 2009). Toshihiko Fukui, a former president of the Bank of Japan, was designated as the chair. Fukui was an expert on the Japanese economy but not on climate change. The majority of the members were concerned with the burdens that could negatively affect Japanese economic activity, when, at the time, Japan was being affected by the worldwide economic downturn that occurred after the bankruptcy of Lehman Brothers. The committee invited four major energy-modeling research groups to calculate emission reduction potentials and the associated economic costs. Clearly, the economic burden of climate change mitigation policy was the most important topic for the committee.

The four modeling groups were as follows. The Research Institute of Innovative Technology for the Earth (RITE) had developed an energy-economic model called DNE21+. It was a bottom-up model that was able to make a comprehensive assessment of technology availability and the associated costs. The Institute of Energy Economics, Japan (IEEJ), had developed the EMDC/IEEJ model, which consisted of a macroeconomic model and various energy demand and supply models. The National Institute for Environmental Studies (NIES) had developed the Asia-Pacific Integrated Model (AIM). It was made up of a group of independent models, including those that assessed the impacts of climate change, bottom-up technology models, and top-down computable general equilibrium (CGE) models. Finally, the Japan Center for Economic Research model was a macroeconomic model that focused mainly on the country's economic trends.

Although the four modeling teams consisted of experts, they were not perceived to be politically neutral. In one way or another, economic-energy models produced different results, even when all of the inputs and underlying conditions were the same. In addition, the committee members had to decide which preconditions and assumptions should be chosen when conducting the modeling. For example, assumptions of future oil prices affect the outcome of modeling calculations of the economic cost of achieving given levels of CO_2 emission reduction, because increases in the oil price cause a decrease in demand for energy, even in the absence of any climate change mitigation measures; thus, a higher oil price will lead to the interpretation of lower costs for the mitigation. Similarly, the time required to recoup the original investment was also important for investors in renewable energy. To recoup initial investment in solar photovoltaic and wind-power facilities typically requires more than ten years. If the assumption used in a model is that investors expect investment to be recouped within three to five years, a very small amount of renewable energy will be introduced under a free market.

Members of the committee were interested to know information concerning marginal abatement cost (MAC) of emission reduction. MAC can be defined as economic cost required to reduce additional unit of emission. In theory, in a free market world, emission reduction actions are taken where the reduction can be achieved at least cost. Once all the low-hanging fruits are gone, the actions will eventually be taken in areas where higher costs are required. Thus, MACs often rise steeply as more GHG emission reduction is required. The committee members, as well as other stakeholders related to Japanese economic sector, believed that equalizing MACs was the most equitable way of comparing emission reduction targets across countries. Thus, calculations by the modeling teams exclusively focused on MACs in each country. The modeling results of the four groups are summarized in Table 4.3. The four assumptions were provided by the committee members as four possible ways to determine Japan's emission reduction target. Both the EU and the United States had already indicated their emission reduction goals for the year 2020 by then, so the committee wanted to set Japan's emission reduction target at a level where the MACs for all developed countries would be equalized.

Table 4.3 Results of models on the economic costs associated with emissions reduction

Assumptions	Marginal abatement cost (MAC)[b]	Base year	Range of outputs by four models on emission target for 2020 relative to the base year (%)			
			Japan	United States	European Union	Annex I total
All Annex I countries set emission reduction targets with MACs equal to that of the EU target (-16% from 1990)[a]	US$48 to $49	2005	-2 to -7	-9 to -14	-10 to -11	-8 to -11
		1990	+2 to +4	0 to +5	-16	-10 to -15
All countries set emission reduction targets with MACs equal to that of the US target (0% from 1990)	US$47 to $62	2005	-2 to -8	-12 to -14	-10 to -12	-10 to -11
		1990	0 to +4	0	-16 to -17	-12 to -15
Annex I countries jointly reduce emissions by 25%, with equal MACs	US$88 to $166	2005	-6 to -12	-30 to -33	-18 to -23	-22 to -23
		1990	+1 to -5	-19 to -24	-23 to -27	-25
Annex I countries jointly reduce 25%, with burden sharing through equal cost per GDP (%)	0.4% to 1.0%	2005	-13 to -23	-19 to -28	-25 to -27	-22 to -23
		1990	-8 to -17	-7 to -18	-30 to -31	-25

[a] The EU target is originally -20% from 1990, but -16% was used in this calculation because the target included 4% by acquisition of credits from the use of carbon markets.
[b] MAC values are the range of results from the four modeling teams.

Source: Prime Minister's Office (2009a)

When Table 4.3 was distributed at the meeting, it was accompanied by two bullet points emphasized in bold:

- Japan's emission reduction rates decrease when compared by marginal abatement cost (MAC), because relatively more emission mitigation technologies are already in place in Japan than in other Annex I countries.

- MAC is the criterion most commonly used to compare emission targets by model analyses, but there are other criteria, such as mitigation cost per unit GDP and emission per capital, for comparison from an equity perspective.

Some experts and environmental NGOs criticized the bullet points associated with Table 4.3 for not being central issues in the negotiations under the UNFCCC (Kiko Network 2009). Particularly, the second point was not entirely accurate, because many researchers had calculated emission allocation rules according to other criteria of responsibility, such as emission per capita or cumulative emission per capita, as well as by using criteria of capacity (such as GDP per capita) as indicators. In Japan, however, economic efficiency measured by MAC was believed to be the most commonly used criterion to compare emission reduction targets across countries from an equity perspective, and some government officials explained to political leaders and to members of councils that they were correct.

There were almost no debates on the consequence of Japan's emission reduction target as it related to the impacts of climate change, including the degree of emission reduction needed globally to reach the long-term 2 °C goal. The Fourth Assessment Report of the IPCC (IPCC AR4) presented a table that showed a range of differences between emissions in 1990 and emission allowances in 2020/2050 for various GHG concentration levels for Annex I and non–Annex I countries as a group (IPCC 2007: Box 13.7). The table showed the required emission reductions based on different principles of equity; Annex I countries were allocated reduction ranges of 25%–40% by 2020 and 80%–95% by 2050 for the 450 ppm-CO_2-equivalent scenario. Although this was not meant to be perceived as the scientific consensus that industrialized countries *should* reduce their emissions by 25%–40% by 2020, it was interpreted in that manner on many occasions. The 25% figure that arose in many Japanese discussions most likely originated from the IPCC report. At the same time, many stakeholders in Japanese decision-making asserted that Japan did not have to take the global impact of climate change into account when deciding on its emission reduction target, because Japan was responsible for only a mere 4% of total global emissions. They argued that any ambitious GHG emission reduction in Japan would have only a small effect on global emissions, and that large emitting countries such as the United States and China were the ones that really needed to reduce emissions.

The majority of the committee members were reluctant to choose assumptions that would favor more ambitious emission reductions, and by using the selected preconditions and assumptions the four modeling teams evaluated six options for emission reduction targets for 2020. These six options were publicized, side by side with the economic costs calculated by the models that would be required to achieve the options (Figure 4.1).

Staff from the Prime Minister's Office held regional meetings to explain the six options, and public comment was accepted between 17 April and 16 May (Prime Minister's Office 2009a). In this type of case, public opinion polls rarely reflect the views of average people, because many of the people responding are considered to be either working for energy-related industries or somehow associated

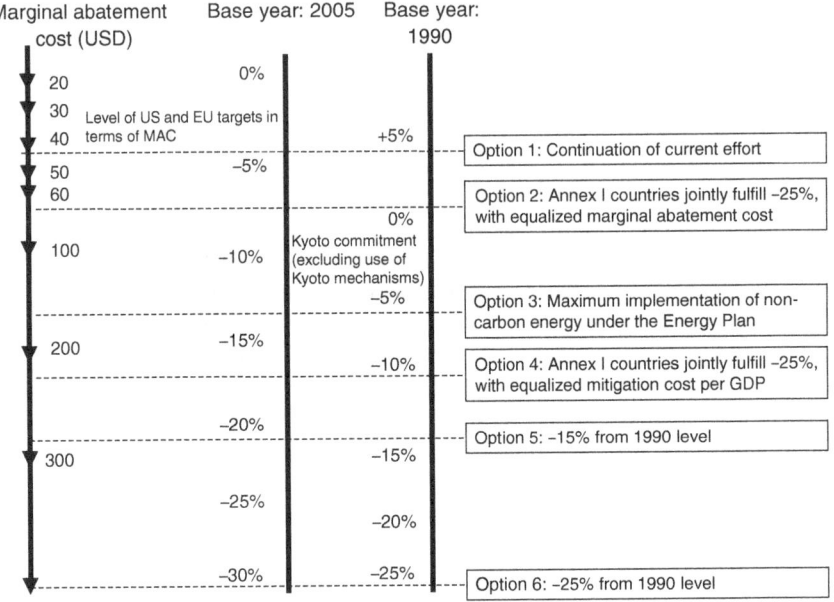

Figure 4.1 Six options for Japan's emission reduction targets for 2020

Source: Prime Minister's Office (2009a)

with environmental NGOs. The former group of people supported Option 1 (-4% from 2005; 4% increase from 1990), whereas the latter group of people supported Option 6 (25% reduction from 1990; 30% reduction from 2005). That said, the Prime Minister's Office came to a conclusion that a compromising option between the two sides would be Option 3 – i.e. 7% reduction from 1990, equal to a 14% reduction from 2005. The prime minister agreed to take this option, and he added one extra percentage point to look environmentally aware.

In June 2009, Prime Minister Aso determined that Japan's midterm target for the year 2020 would be a 15% reduction from a base year of 2005. This target was actually equal to a 7% reduction from the 1990 level, as Japan's GHG emission increased nearly 8% between 1990 and 2005. Because the emission reduction target under the Kyoto Protocol for the years between 2008 and 2012 was a 6% reduction from the 1990 level, this target meant that only an additional 1~2% reduction would be made between 2012 and 2020. Prime Minister Aso emphasized that this target was set for domestic emissions only, and that further reductions could be made possible by adding sequestration by sinks and through the acquisition of emission allowances or credits from abroad. Still, the decision was criticized by environmental NGOs as being too weak to tackle climate change.

The Aso administration was also criticized by environmental NGOs for other policies he had agreed upon independent of the climate agenda. In March 2009, the coalition government of the LDP and Komei Party agreed to lower the price of highway tolls to stimulate economic activity. This policy increased the incentives for people to use private cars rather than public transportation, thus potentially increasing CO_2 emissions.

Another policy introduced during the Aso administration was the use of "eco-points" in purchasing household electric appliances, such as refrigerators, TV sets, and air-conditioners. Consumers who purchased appliances labeled as the most energy efficient would be given eco-points, which could be used to receive discounts on other purchases. The Aso administration stressed that this policy was intended to stimulate consumption of environmentally friendly products, but environmental NGOs opposed the policy because people who had never used air-conditioners before started buying them. Similarly, in 2009, a tax deduction was offered on the purchase of relatively more energy-efficient cars (eco-cars). These economic incentives were effective in changing people's consumption patterns toward the purchase of more energy-efficient products, but the effect on total GHG emission was uncertain. In any case, many similar policy instruments geared toward stimulating economic activity were approved under the Aso administration.

Heads of states gathered at the L'Aquila G8 Summit in Italy in July 2009, and they were anxious to discuss the worldwide economic and financial crisis that had begun in 2008. The outcome of the summit focused on the urgency of taking action to move toward economic recovery. The participants also took note of the G8's significant role in directing the world toward achieving a successful outcome at the upcoming COP15 meeting in Copenhagen, which was to take place in five months. The group affirmed its determination to shape a global and comprehensive post-2012 agreement by the end of 2009 in Copenhagen. It reaffirmed, by referring to the IPCC AR4, that they understood that the global average temperature should not exceed preindustrial levels by more than 2 °C. To achieve this level, the leaders showed their willingness

> to share with all countries the goal of achieving at least a 50% reduction of global emissions by 2050, recognizing that this implies that global emissions need to peak as soon as possible and decline thereafter. As part of this, we also support a goal of developed countries reducing emissions of GHGs in aggregate by 80% or more by 2050 compared to 1990 or more recent years. Consistent with this ambitious long-term objective, we will undertake robust aggregate and individual mid-term reductions, taking into account that baselines may vary and that efforts need to be comparable. Similarly, major emerging economies need to undertake quantifiable actions to collectively reduce emissions significantly below business-as-usual by a specified year.
>
> (L'Aquila G8 Summit 2008)

As other leaders of the G8 member countries returned home to work on these firm commitments, Prime Minister Aso had no chance to do so. In July 2009, the LDP experienced a great loss in the Diet Lower House election, after which Aso resigned.

The DPJ's Hatoyama administration

The July 2009 election resulted in a landslide victory for the Democratic Party of Japan (DPJ), and Yukio Hatoyama became the new prime minister in September 2009. During the election campaign, the DPJ showed its willingness to act positively toward taking action on climate change. In his speech at the UN special summit on climate change, held in New York, Hatoyama announced that Japan would change its midterm emission target for 2020 to a 25% reduction from the 1990 level. This target was considered to include not only emission reductions at the domestic level but also sequestration by sinks and full utilization of carbon markets. He also announced the "Hatoyama Initiative," a plan to support developing countries financially and technically to mitigate climate change, as well as to support these countries in their capacity to report their activities.

His new ambitious emission target was welcomed by many internationally, but the main problem with Hatoyama's new emission reduction target was that it had not previously been discussed inside the Japanese government. Even Hatoyama himself did not have a firm idea of how Japan could achieve this goal. Upon returning from New York, Hatoyama called for the establishment of a new Task Force on Global Warming under the cabinet and instructed it to determine how a 25% reduction target could be achieved, how much it would cost, and the economic implications for Japanese economy. The task force met six times between October and December 2009. The group was under a time constraint because it needed to finalize its work before COP15, so the same modeling groups that participated in the previous consultation under the Aso administration were again asked to make estimations, but using different assumptions this time. For example, the previous models did not incorporate any potential economic benefits accrued by investments in low-carbon technologies. The Hatoyama administration's aim was to include these types of positive impacts of emission reduction targets in the estimates. The task force also considered the amounts of emission allowances and credits that could be purchased abroad.

The task force published an interim report in November 2009 (Prime Minister's Office 2009b). It was not able to change assumptions on the prospects of future levels of productivity, because these estimates were based on each manufacturing company's production plans and had to be used. The models showed that the 25% emission reduction target was achievable at the least cost when most of the reductions were fulfilled by purchasing emission allowances from abroad. Although this conclusion seemed reasonable, it did not necessarily mean it was the best way to achieve the 25% emission reduction target for the Japanese economy in the long run, because Japan would almost certainly have to aim for additional reductions in the ensuing years after 2020, and thus it would be best for Japan to invest in Japan on low-carbon technologies. Some of the emission reduction effort needed to be made domestically, even if some of the measures were more costly initially than purchasing carbon credits.

In the fall of 2009, an increasing number of countries began to introduce their emission pledges for 2020. COP15 was to be held in Copenhagen in

December 2009, and the Danish government requested a high-level meeting – at the head-of-state level. Before this, the COP had held ministerial high-level meetings, but this was the first time that a COP of the UNFCCC had invited heads of state to discuss climate change. In many circles, this was seen as a message that countries should come to Copenhagen with ambitious emission reduction targets.

In the United States, which had been under the Obama administration since January 2009, the environment on climate change legislation had changed. The *American Clean Energy and Security Act of 2009*, also known as the Waxman-Markey Bill, passed the House of Representatives on 26 June 2009. The bill proposed a cap-and-trade system for the 2012–2050 period. The bill required a 17% emissions reduction from 2005 levels by 2020 and aimed at an approximate 83% reduction by 2050. Although this bill was later abandoned by the Senate in July 2010, these targets were the emission reduction targets proposed by the United States during the Copenhagen meeting.

China, which had surpassed the United States to become the world's largest GHG emitter by 2008, also started to become more proactive in negotiations on climate change. At the UN climate summit in September 2009, Hu Jintao, General Secretary of the Communist Party of China, publicized China's willingness to reduce its emissions per GDP significantly from its 2005 level, to increase the share of non–fossil-fuel energy use to 15% of total energy use by 2020, and to increase its forest cover by 40 million ha from 2005 levels. Two months later, China additionally announced that its emission reduction target for 2020 would be to reduce CO_2 emissions per unit of GDP by 40%–45% as compared with the base year of 2005.

Indian Environment and Forests Minister Jairam Ramesh was more interested in the establishment of the Green Climate Fund (GCF), which would be set up to be independent of the World Bank. India also pledged a 20%–25% decline in carbon intensity (emission per GDP) below 2005 levels.

COP15 was held from 7–19 December. It was supposed to be the final opportunity to respond to the preliminary steps taken by the Bali Action Plan to arrive at a global climate treaty. The expectation of arriving at a binding agreement appeared to be little more than a dream a month before COP15 at the Barcelona meeting, where negotiators confirmed that too little progress had been made to reach the original aim. Japan was not a hardliner in terms of insisting on the originally planned outcome of COP15. The bottom line of the Japanese government was to see all major countries, including the United States and China, be included in a same agreement. Environment Minister Sakihito Ozawa gave a speech in which he pledged JPY1,750 billion (about USD17 billion) over three years to tackle climate change in developing countries as a part of the Hatoyama Initiative. Hatoyama also attended the head-of-state meeting, and he was happy to know his ambitious emission reduction target of 25% was welcomed by other countries.

Negotiations faced a deadlock, not knowing how to resolve the many conflicting agendas on the table. More than 110 heads of state arrived in Copenhagen on the second week of COP15 to address the issue. The last two days of the high-level segment of the meeting were dedicated to crafting a political declaration by the presidents and prime ministers themselves (Egenhofer and Georgiev 2009). The result was a political declaration known as the Copenhagen Accord; it was not adopted but only taken note of by the COP, because of strong objections from several countries that were not involved in the small drafting group, which consisted of about two dozen heads of state (UNFCCC 2009).

The main features of the accord were as follows:

- Recognition of the scientific view that the increase in global temperature should be limited to less than 2 °C. To reach this long-term goal, cooperation is needed to achieve the peaking of global and national emissions as soon as possible, recognizing that the timeframe for peaking will be longer in developing countries and bearing in mind there will be social and economic development and poverty implications.
- Annex I parties commit to implement individually or jointly quantified economy-wide emissions targets for 2020, to be submitted by the Annex I parties to the secretariat by 31 January 2010. Delivery of reductions and financing by developed countries will be measured, reported, and verified in accordance with existing and future guidelines adopted by the COP.
- Non–Annex I parties to the UNFCCC will implement mitigation actions, including those to be submitted to the secretariat by non–Annex I parties by 31 January 2010. Mitigation actions taken by non–Annex I parties will be subject to their domestic measurement, reporting, and verification, the results of which will be reported through their national communications every two years. Nationally appropriate mitigation actions that are supported internationally will be subject to international measurement, reporting, and verification in accordance with guidelines adopted by the COP.
- The crucial role of reducing emissions from deforestation and forest degradation will be recognized, as will the need to enhance removal of GHG emissions by forests and the need to agree to provide positive incentives for such actions through REDD-plus (reducing emissions from deforestation and forest degradation in developing countries).
- Adaptation to the adverse effects of climate change and the potential impacts of response measures is a challenge faced by all countries.
- The collective commitment on finance by developed countries is to provide new and additional resources approaching USD30 billion for the period 2010–2012, with allocation balanced between adaptation and mitigation. Developed countries commit to a goal of jointly mobilizing USD100 billion a year by 2020 to address the needs of developing countries. This funding will come from a wide variety of sources, both public and private and bilateral and multilateral, including alternative sources of financing. A significant portion of such funding should flow through the Copenhagen GCF.

The paragraph on GCF was of great concern to Japan, and particularly to MOFA and MOF. USD100 billion was an incredible amount compared with conventional funding as overseas development aid (ODA). Still, it was possible for Japan to agree to this commitment because there was a sentence that allowed countries to count a wide variety of financial resources, including those by private firms, and also because the United States could also accept the commitments in this paragraph.

In January 2010, Japan communicated that its pledged 2020 emission reduction target was a 25% reduction from the 1990 level; this was premised on the establishment of a fair and effective international framework in which all major economies would participate and on the agreement by those economies on ambitious targets. As this statement made clear, the 25% reduction target was effective only when other major countries also agreed on ambitious targets. The communication did not indicate how the target might change if this precondition were not met.

In early 2010, after COP15 was over, a new team was set up by the MOE to develop a road map to achieve the 25% reduction target. The Road Map Consultation Committee consisted of more than 50 researchers and experts involved in different working groups, such as the automobile, building, regional development, agriculture, and energy working groups. In March, Environment Minister Ozawa published the "Road Map Towards Achieving the Emission Reduction Target by 2020" as his personal proposal. This report did not have official status as Japan's decision because he had not consulted with the other relevant ministries when preparing the statement. Nevertheless, in airing this report as the minister's personal view, the minister was expressing the view of the Environment Ministry to the public without having it watered down by opponents' criticism or compromise. The report emphasized that a 25% emission reduction by 2020 and an 80% reduction by 2050 were achievable without harming the Japanese economy. The analyses in the report indicated the need to improve energy efficiency throughout all sectors of the economy. Zero-emission buildings and houses, less-carbon-intensive automobiles, and alternative means of transportation were considered critical to meeting the goals. In addition, investments in these types of infrastructure could be economically beneficial to the Japanese economy. The road map also suggested the construction of eight new nuclear power plants by 2020. The plan suggested that policy instruments such as cap-and-trade emissions trading and a carbon tax would be effective in motivating society to move in the direction of developing low-carbon technologies.

Meanwhile, the MOE was attempting to establish new legislation specifically addressing climate change policies and measures, tentatively calling it the "Basic Law concerning Global Warming Measures." The major purpose of creating a new law was to explicitly mention the global long-term 2 °C target and also Japan's 25% emission reduction target by 2020. In addition, it was important to emphasize that this target was approved by the DPJ, because all the other legislation and guidelines on climate change had been adopted under LDP administrations, and none of them had much legal power. The ministry invited comments on the proposed law, which was widely opposed by industry groups and employees of energy-intensive industries. In general, the collected comments asserted that a 25% emission reduction target might result in adverse effects on Japanese economic growth, and people were generally opposed to the introduction of a carbon tax and a domestic ETS. The MOE had to withdraw its proposed bill.

The DPJ also requested that the Agency for Natural Resources and Energy review its *Basic Energy Plan*, because use of energy in the future had to be consistent with the CO_2 emission reductions established in the environment minister's road map.

COP16 was held in Cancun, Mexico, in December 2010. The most important aim of the meeting was to transform the Copenhagen Accord into a package of COP decisions to anchor the political declaration into official procedural documents. On the first day of COP16, one of the Japanese delegations from METI confirmed that Japan's emission reduction target for 2020 was pledged as a target under the Copenhagen Accord, but that Japan did not intend to scribe the target in the Annexes of the Kyoto Protocol under any circumstances or preconditions. This was a strong message that left no room for negotiation. Although many stakeholders had already recognized that Japan's 2020 target was a pledge and not a legally binding cap, others – particularly environmental NGOs – criticized Japan for sending this strong message on the very first day of the COP.

After 2 weeks of intensive discussion, the meeting successfully adopted the Cancun Agreement, which was a means of adopting the contents of the Copenhagen Accord into COP decisions. The agreement formerly accepted the long-term goal as the global 2 °C limit. The pledges made by countries on the mitigation efforts though 2020 were incorporated into the agreement. The countries agreed to submit biannual reports to check on progress toward achieving the 2020 targets. The GCF was also approved.

Involvement of political leaders

Koizumi's administration lasted until September 2006. After he agreed on the Marrakesh Accord in 2001, he was basically supportive of Japan's ratification the Kyoto Protocol and the goal of a new round of negotiations for the post-2012 period. It would be more honest, however, to say that he did not care much about the process, because he had other business to attend to. Inside Japan, Koizumi's priority was privatization of the postal service, which had been run by the government for more than a century. The service employed more than 200,000 workers, most of whom supported the LDP. Thus, privatization of the postal service was considered to be politically difficult. Koizumi urged the Diet to pass the relevant legislation in 2005, and the postal service became privatized in 2007.

Another issue in Japan at the time was the overwhelming budget deficit. Koizumi minimized expenditures by decreasing pension coverage and medical care. He also cut the budget for public construction works, which had been protected by construction-*zoku* ("clan" or group of supporters) from the LDP. Furthermore, he decreased the amount of ODA. MOFA was concerned that Japan's international presence would decline, but improving the budget deficit was considered the top priority.

In terms of foreign policy, as did his predecessors, Koizumi cherished good relations with the United States, but, in August 2004, a helicopter crashed onto a university campus in the neighborhood of the Futenma US military base in Okinawa prefecture. Luckily, no one was hurt, because the accident occurred during the summer break, but residents' opposition to the US military base in Okinawa became much stronger. After nearly two years of discussions with the Mayor of Nago, where Henoko is located, the Japanese government was finally able to obtain the city's acceptance of building a new US military base there. Koizumi also continued dialog with North Korea to return Japanese abductees. Koizumi's relatively long six-year term enabled him to make changes both inside and outside Japan. In terms of climate change, however, there appear to have been few occasions during the Koizumi administration when government officials needed political leaders to mediate conflict among themselves.

The summer of 2005 marked the sixtieth anniversary of the end of World War II. As Prime Minister Murayama had done a decade ago, Koizumi was expected to make a *danwa* (discourse) regarding the war. In his speech, he stated, "Sincerely facing these facts of history, I once again express my feelings of deep remorse and heartfelt apology, and also express the feelings of mourning for all victims, both at home and abroad, in the war." By using the term "feelings of deep remorse and heartfelt apology," Koizumi was trying to improve Japan's relationship with China and Korea, but this was difficult because he had already offended these countries by visiting Yasukuni Shrine, a war memorial for Japanese soldiers, several times during his administration.

Even though Koizumi had to address a wide variety of issues during his administration, he always appointed environmentally aware people as environment ministers. Yoriko Kawaguchi continued in the post after the Mori administration, and she helped pave the way for the Kyoto Protocol's ratification. The next Environment Minister, Hiroshi Ohki, was the president of COP3 and personally supported the Kyoto Protocol. Environment Minister Shunichi Suzuki succeeded Ohki and worked hard to reform the earmarked oil special account. The next environment minister, Yuriko Koike, was also popular. She served in office for three years and was responsible for ideas such as the Cool Biz campaign.

A series of short administrations followed Koizumi's relatively long run in office. Prime Minister Shinto Abe served one year between September 2006 and September 2007. In an attempt to mend the tarnished relationship with China and Korea under the Koizumi administration, Abe visited Beijing and Seoul to attend bilateral summit meetings. An especially contentious issue between Korea and Japan was that of comfort women during World War II. At the same time, Abe proposed the catch phrase, "Beautiful Country," in which he expressed his strong willingness to change the interpretation of Article 9 of the Japanese Constitution so that Japanese defense troops could be sent overseas in some cases. His "Beautiful Planet" concept, presented in May 2007, was drafted as an extension of his Beautiful Country idea.

Abe also thought a great deal of the upcoming G8 Summit to be hosted by the Japanese government. He chose Toyako, a remote lake resort in Hokkaido in northern Japan. He actually visited the venue, because he expected to be there at the time of the meeting in July 2008. Unfortunately, Abe's administration did not last long enough for him to host the Toyako G8 Summit. Many scandals and political conflicts had emerged among LPD Diet members. Although Abe himself was not targeted by the scandals, the LDP lost credibility with voters and lost an election in the Upper House in June 2007. The DPJ won a landslide victory in the Upper House and started making objections to bills proposed by the LDP. Abe took the responsibility for LDP's loss and resigned his position three months later. There were also rumors that Abe had had health problems.

The next prime minister, Yasuo Fukuda, was in the position for a year, beginning in September 2007. The most critical issue in the realm of foreign policy at that time was antiterrorism activities, which had been agreed upon under the Koizumi Administration just after the 9/11 incident in the United States in 2001. The legislation allowed the Japanese government to send troops overseas under certain conditions. The legislation had a deadline for the allowance, however, and the Diet had to discuss continuation of the bill. There had always been strong voices inside Japan opposing Japan's acquisition of military power, in fear it might follow the same path that led to the country's involvement in World War II. Meanwhile, the United States pressured the Japanese government to make sufficient contributions in the field of security and antiterrorism. Prime Minister Fukuda, the LDP leader, was unable to lead the discussion in the Diet, because the DPJ's leader, Ichiro Ozawa, was exerting increased influence over Japan's politics.

An important domestic issue of the time was tax reform. Since the Koizumi Administration, the MOE had targeted the oil special account, which was funded by an earmarked tax on gasoline and used specifically for the purposes of road construction and maintenance. A partial reform was made during the Koizumi administration by Environment Minister Suzuki, but there were influential road-*zoku* LDP members who objected to completely abolishing the account. Fukuda basically agreed with Koizumi's position that the tax revenue should be used for other purposes, such as environment-related activities. The DPJ, however, emphasized that the price of gasoline should be lowered and recommended complete abolishment of the tax. The debate reached no conclusion. Fukuda and MOF's idea was preferred by those wishing to promote climate change mitigation policies, but there were too many opponents to this agenda.

Fukuda faced a great deal of political difficulty domestically, but he was popular outside Japan. He cultivated a good relationship with China's president, Hu Jintao. Fukuda was invited to the opening ceremony of the Beijing Olympic Games. He was eloquent in discussions with various political leaders overseas. In July 2008, Fukuda hosted the G8 Toyako Summit – his first time attending a G8 meeting. The top agenda of the meeting was climate change, and Fukuda was well prepared for it. He gained support from US president G.W. Bush and other members of the summit, and, most importantly, he succeeded in concluding the meeting with the long-term global emission target of halving emissions by 2050. The G8 Summit in Toyako was a significant event for Fukuda, not only in terms of climate change. This was the first time Fukuda met Russian president Dmitry Medvedev, and they issued a joint statement calling for a resolution of the nearby Northern Territories dispute, which had been an issue, at least in Japan, since the end of World War II (Hook et al. 2012).

Despite his success in taking a leadership role at the G8 Summit, public opinion polls showed decreasing support for the LDP, while at the same time the DPJ was increasing its power. Fukuda decided to resign his post immediately after the summit and asked Taro Aso, also an LDP member, to take over the position.

Aso began his term as Japan's prime minister in September 2008 but served for less than a year. Unfortunately for Aso, the bankruptcy of Lehman Brothers occurred just after his inauguration. The world went into a deep recession, but Aso was given little opportunity to take meaningful actions to aid the economic recovery. At this same time the Japanese government was discussing the emission reduction target for 2020, so Aso was not willing to commit to any plan that could lead to additional burdens on Japan's lagging economy.

The environment minister under the Fukuda and Aso administrations was Tetsuo Saito. He was from the Komei Party, which had formed a coalition government with the LDP during that period. To maintain a strong coalition the LDP needed to have someone from the Komei Party in the cabinet, and for the LDP the position of environment minister was considered one of the least significant. Saito's educational background was in physics; this meant that he was capable of dealing with the scientific issues, but he had little influence in national decision-making on emission reduction targets.

The Aso administration was not very popular among the Japanese people, and the election of Lower House Diet members in August 2009 resulted in a landslide victory for the DPJ. Yukio Hatoyama became the first prime minister from the DPJ in September 2009. Although the DPJ members were excited to take over as the ruling party, they were not particularly well prepared to actually take the lead in Japan's policy-making.

Just after Hatoyama became prime minister, he participated in the climate summit held by the United Nations Headquarters. He announced Japan's ambitious emission reduction commitment of 25% from the 1990 level by 2020. This target was said to have been recommended by Tetsuro Fukuyama, also a DPJ member, who was personally involved with global environmental issues even before the DPJ won the election and who recommended inclusion of a 25% reduction target in DPJ's manifesto during the election campaign. As noted previously, this was an ambitious target, but it had not been sufficiently discussed inside the Japanese government. Furthermore, it came as a great surprise for Japanese industries and METI, and they questioned how Hatoyama planned to achieve such an ambitious target.

Hatoyama surprised both Japanese and US government officials when he insisted on moving the US army base in Okinawa Prefecture to another prefecture in Japan. Previously in 2006, the US government and LDP prime ministers had agreed that the base would be moved from Futenma to Henoko, both within Okinawa Prefecture. Okinawa residents had strongly opposed this plan but (up to this point) to no avail. They hoped that Hatoyama could change the agreement, but the suggested change was not well received by the Obama administration, and the relationship between the United States and Japan became strained. Entering into 2010, Hatoyama was pressured by other political leaders in Japan, as well as from the United States, to affirm that the 2006 agreement between the two countries was a confirmed final agreement. Hatoyama therefore began to back down on the issue. Meanwhile, the Social Democratic Party of Japan (SDPJ) had been in a coalition position with the DPJ and thus shared the ruling government's authority. The SDPJ position – unchanged since before joining the coalition – was to do away with the US military base on Okinawa. Thus, Hatoyama's shift on this issue essentially meant dissolution of the coalition. Furthermore, Ichiro Ozawa, the most politically powerful politician of the time in the DPJ – widely viewed as the man who "sat behind the curtain" and controlled Hatoyama – was criticized for receiving donations from private companies and using the money to purchase land to build an office. Popular support for the DPJ decreased rapidly. Hatoyama and Ozawa both decided not to run in the next election for prime minister. The DPJ leaders then internally chose Naoto Kan as the next head of the DPJ and consequently the next prime minister.

Involvement of nonstate actors

Industries

The Kyoto Protocol was very unpopular among Japanese industries. They disliked the protocol because (1) they thought the 6% reduction target was harmful to the Japanese economy; (2) the United States, the largest trading partner of Japan, withdrew from it, not only making the Kyoto Protocol ineffective in terms of reducing global GHG emissions but also putting Japanese industries in an even less competitive position against US companies; and (3) they could see Chinese industries rapidly catching up to developed industrialized countries, and this made them more concerned about both Japan's own internal economic growth and regional energy security.

Thus, in November 2003, Keizai Doyukai, a leading business group, commented on a report issued by the Global Warming Mitigation Tax Committee organized under the Central Environment Council (Central Environment Council 2004b) in opposition to any taxes that were not well designed. Also in April 2005, Keizai Doyukai published a comment on the Kyoto Protocol Target Achievement Plan. The comment had eight main pillars: national and local governments should take action first; monitoring of emissions is necessary; voluntary actions by industries should be prioritized; the CDM and other Kyoto Mechanisms should be utilized; no new taxes should be imposed; emissions trading should not be implemented domestically unless the scheme is proven not to harm economic growth; technology development should be aimed at the long term; and government should take the leadership role in international contributions. This strong opposition to economic instruments such as taxes and emissions trading was influential enough to get the introduction of effective emission reduction policies removed from the Kyoto Protocol Target Achievement Plan.

By 2004 or 2005, industry groups were well prepared to insist on the abolishment of the Kyoto Protocol. In the EU, industries basically accepted the EU's regional ETS and were more interested in the introduction of border adjustment measures to maintain their international competitiveness and to avoid carbon leakage (Dröge 2009). Japanese industries, on the other hand, preferred to see carbon leakages be dealt with in multilateral agreements, with meaningful participation by all major emitters. This meant that the post-2012 multilateral regime should include commitments for major new emerging economies so that leakages would not occur.

At the time of the Bali Action Plan in 2007, the Keidanren strongly insisted on the sectoral approach – an approach, by its own definition, that meant that industry sectors would develop their own voluntary emission targets and that national absolute emission reduction targets would be unnecessary. In the ensuing years (2008–2009), when Japan was holding internal discussions on the midterm emission reduction targets for 2020, industry groups pressured politicians and government agencies to aim at less ambitious emission reduction targets.

Labor unions

Rengo (the Japanese Trade Union Confederation) is Japan's largest labor union organization, with nearly 7 million members. It basically speaks on behalf of employees and workers. Whereas the LDP was strongly supported by Keidanren, other industry groups, and groups of employers, the DPJ was supported mainly by Rengo. Rengo was happy to see the DPJ take the majority and be in the cabinet in 2009, but the employees' position was no more climate friendly than that of the industry groups. Labor unions also preferred policy instruments that led to more economic activity, more jobs, and less economic burden on workers. Although basically supportive of green issues, Rengo was also against the introduction of environmental taxes and supportive of abolishing highway tolls.

One of the powerful labor unions in Rengo was that of workers at nuclear power plants. One of the reasons why the DPJ was able to propose an ambitious emission reduction target was that, within the discussion of the emission reduction target, there was the hope that Rengo members would be employed to build more nuclear power plants in Japan.

Local governments

Some prefectural governments were far more engaged in climate change policies than was the national government. That was especially true of the Tokyo Metropolitan Government (TMG) during this period. Traditionally, the TMG was unique in the sense that it had great administrative power over the largest megacity in Japan, covering about 10% of the entire Japanese population. Shintaro Ishihara, governor of Tokyo Prefecture, was in office from April 1999 to December 2012. This was much longer than the relatively short administrations of prime ministers at the national level. Under his leadership, environmental policies were popular among Tokyo citizens.

In June 2008, the TMG announced that it would start its own municipal level ETS in 2010, but initial discussions of the introduction of the cap-and-trade scheme had started much earlier. In 2000, the TMG amended the *Tokyo Metropolitan Pollution Prevention Ordinance*, replacing it with the *Tokyo Metropolitan Environmental Security Ordinance*. It also established the Tokyo Carbon Reduction Reporting Program, which required large emitting facilities to report their GHG emissions and actions and encouraged them to set voluntary emission reduction targets, beginning in 2002. With the new reporting requirement, the TMG was able to obtain data regarding GHG emissions by facilities in Tokyo. This was helpful when it later began to design a cap-and-trade-type ETS.

In 2005, the TMG amended the *Tokyo Metropolitan Environmental Security Ordinance* once again. The amendment introduced a mechanism to provide guidance and advice to building owners on the GHG emission reduction plan. The ordinance also enabled the TMG to evaluate companies' plans, give awards to outstanding facilities, and make emission reduction plans open to the public. The Bureau of Environment of the TMG assessed reported emissions and actions, and this again served as a solid basis for designing the ETS. The assessment work was critically important in that it allowed the TMG to study, through site visits at about 1,000 installations, reduction potentials and the concerns of operators and owners of installations.

In June 2007, the TMG announced its Tokyo Climate Change Strategy, which proposed the introduction of a mandatory emissions reduction program for large facilities. On the basis of a year-long deliberation at the Environmental Council and Stakeholder meetings, in June 2008 Shintaro Ishihara finally submitted a bill to the second regular meeting of the Tokyo Metropolitan Assembly. The bill introduced mandatory targets for reduction in overall GHG emissions for large-scale emitters as part of an ETS. The amended ordinance was enacted in April 2009. Through such preparation processes, confidence and consensus were built over time, eventually leading to successful introduction of the cap-and-trade program in April 2010. The Tokyo cap-and-trade scheme was recognized as the world's first urban program to include office buildings. The scheme targeted about 1,400 facilities with large GHG emissions; 1,100 of these were business facilities and office buildings. These 1,400 facilities accounted for an estimated 40% of CO_2 emissions from the Tokyo metropolitan area.

With guidance from the TMG, Saitama Prefecture also introduced emissions trading in 2011. Some other prefectural governments, such as Kyoto and Shiga Prefectures, were also relatively well known for their forward-looking environmental positions, which were independent of national policies.

Many other local governments were more interested in engaging international cooperation with local governments in developing countries in Asia, even though international cooperation is not an official mandate of local governments. Out of 47 prefectures and some 800 cities in Japan, 43 prefectures and 23 cities have engaged in at least some form of international environmental cooperation activities such as the acceptance of trainees, dispatch of experts, and hosting of intercity network programs for sustainable city management in Asian and other developing countries (Nakamura et al. 2011). Their motivations were based on their desire to make contributions according to their local experience and human capital, promotion of environmental businesses, and response to transboundary pollution issues.

Tackling climate change in Asian countries was effective particularly to improve air quality at regional level.

Scientists

A rapidly increasing amount of research activity related to climate change was under way by the early twenty-first century in Japan. Much of this research was related to IPCC activities and included the authorship of IPCC assessment reports.

As for evaluation of mitigation policies, several modeling groups were heavily involved in governmental decision making. Some studies proved that major emission reduction in Japan is possible without harming economy (Ashina et al. 2012). Other studies focused on comparison of marginal abatement cost curves across countries to emphasize Japan's relative disadvantage to reduce GHG emissions (Akimoto et al. 2012).

Concerning impact of climate change, the Atmospheric Circulation Model was developed to estimate future global warming. Studies showed increasing temperatures at the global level, including in Japan. The MOE funded a study group to draft a report on detailed calculations of the economic costs associated with the adverse effects of climate change (Matsui et al. 2009; Ondanka Eikyo Yosoku Project 2009). Major adverse effects of climate change that are likely to occur in Japan in a near future detailed in the reports were inundation by flooding due to concentrated rainfall, landslides in mountainous areas, climate stress on ecosystems (particularly *buna* [beech] and pine trees), changes (both increase and decrease depending on each region) in the crop yield of rice, loss of sandy beaches, inundation by high tides, and an increasing number of deaths due to heat stress. The study group estimated these impacts of climate change in Japan through 2100 by using three scenarios – a 450-ppm stabilization scenario (a scenario that aims at a 2 °C goal), a 550-ppm stabilization scenario (a less stringent goal to mitigate adverse impact of climate change), and business-as-usual scenario – to compare the costs of the impacts. The report clearly estimated that costs would increase because of the impacts of climate change if the world did not sufficiently respond to climate change. For example, flood damage was estimated to reach JPY8 trillion (about USD70 billion) annually by the end of the twenty-first century.

The Science Council of Japan established the "Investigation Committee on Human-Induced Global Environmental Problems such as Global Warming" in 2007–2008. It consisted of 12 experts in the field of climate change. However, their views were quite diverse. Some insisted that emissions reduction should aim at a 2 °C limit in the long run, whereas others emphasized that the economic costs associated with reducing GHG emissions could be significant, so the 2 °C target should be reconsidered. They were unable to reach a clear consensus on policy instruments or emission targets (Science Council of Japan 2009).

Environmental NGOs

Japanese environmental NGOs dealing with climate change were still growing in number of members and had increased their capacity to create alternative policy recommendations. Meanwhile, the general public's attention on global environmental problems had declined since the Kyoto COP3 meeting. However, it peaked again in 2007, when the IPCC was honored with a Nobel Prize and mass media refocused attention on the topic.

Summary

Unfortunately, many Japanese people – particularly those affiliated with industry – were not really attached to the Kyoto Protocol, a historic multilateral agreement adopted in Japan. Rather than taking a lead in acting to combat climate change, many Japanese stakeholders compared themselves with stakeholders in the United States and China, which had rejected the Protocol: they felt it was unfair to abide by it when other major emitting countries were not.

This feeling of uneasiness toward the Kyoto Protocol affected Japan's position on debates about the post-2012 international institution. Japan's paramount hope was to have major emitters participate in the next round of agreement. Japan repeatedly emphasized this "wide participation" aspect of any future agreement, but it also wanted to become a follower, rather than the world leader, in reducing emissions. It remained careful not to commit to anything ambitious unless other major countries – particularly the United States and China – also did so.

Japan was also generally not that interested in the impacts of climate change. It considered itself to be a small country in terms of emissions, with only 4% of the global total. Japan thought that the adverse impacts of climate change could not be changed by Japan alone, even if it were to reduce its emissions to zero. It was the role of big emitters to make significant emission reductions if the world were to achieve the long-term target of a 2 °C limit on global temperature increase. Even this long-term target was criticized by some industry groups, who claimed that the temperature target had been determined politically rather than scientifically. Some people affiliated with Japan's industries challenged climate change experts by asking, "What would happen if the temperature exceeded the limit by 0.1 °C – would it make a big difference?" These types of arguments framed Japan's internal discussion on climate change, leading to an avoidance of discussions on the dangerous impacts of climate change and the setting of ambitious emission reduction targets. Almost no voices were heard saying that Japan should take the lead in combating climate change by setting an ambitious emission reduction target and demanding that other countries follow suit. The more popular option was to sell energy-efficient Japanese technologies outside Japan to achieve low-carbon societies elsewhere.

Japan's emission reduction target for 2020 was determined at a time when its prime ministers were changing virtually annually. It was very difficult for any of the prime ministers to propose ambitious GHG emission reduction targets when their political position was constantly unstable. Many other pressing agendas had to be taken care of, and primer ministers often had no time, expert knowledge, or political will to prioritize global environmental issues.

The DPJ's victory in 2009 marked a turning point in Japan's political history and its climate-change policy-making. However, this was not entirely a good thing, because DPJ supporters disliked climate change mitigation policies that they perceived as costly. The one option widely supported by the DPJ was the expansion of the use of nuclear power in Japan. Prime Minister Hatoyama's emission reduction target of 25% was praised by many countries, and, despite the failed negotiations at Copenhagen, the COP that followed succeeded in reaching the Cancun Agreement. At the time, Japan was expected to be a frontrunner in achieving one of the most ambitious emission reduction targets in the world – that is, until March 2011.

References

Akimoto, Keigo, Fuminori Sano, Takashi Homma, Kenichi Wada, Miyuki Nagashima, and Junichiro Oda (2012) "Comparison of Marginal Abatement Cost Curves for 2020 and 2030: Longer Perspectives for Effective Global GHG Emission Reduction," *Sustainability Science*, 7, 157–168.

Ashina, Shuichi, Junichi Fujino, Toshihiko Masui, Tomoki Ehara, and Go Hibino (2012) "A Roadmap towards a Low-Carbon Society in Japan using Backcasting Methodology: Feasible Pathways for Achieving an 80% Reduction in CO_2 Emissions by 2050," *Energy Policy*, 41, 584–598.

Azar, Christian (2005) "Post-Kyoto Climate Policy Targets: Costs and Competitiveness Implications," *Climate Policy*, 5(3), 309–328.

Bodansky, Daniel, with contributions from S. Chou and C. Jorge-Tresolini (2004) *International Climate Efforts Beyond 2012: A Survey of Approaches*. Washington, DC: Pew Center on Global Climate Change.

Bosi, M. and J. Ellis (2005) *Exploring Options for "Sectoral Crediting Mechanisms"*. Paris: OECD/IEA.

Bradley, Richard and Kevin Baumert (2005) *Growing in the Greenhouse – Protecting the Climate by Putting Development First*. Washington, DC: World Resources Institute (WRI).

Central Environment Council (2004a) *Climate Regime Beyond 2012: Basic Considerations*, Interim report. Global Environment Committee, Central Environment Council. Tokyo: MOE.

Central Environment Council (2004b) *Interim Report on Taxation for Global Warming Mitigation and Related Polices and Measures*. Available online at:http://www.env.go.jp/policy/report/h16–02/index.html (accessed 15 September 2015).

Clémençon, Raymond (2008) "The Bali Road Map: A First Step on the Difficult Journey to a Post-Kyoto Protocol Agreement," *Journal of Environment and Development*, 17(1), 70–94.

Council on the Global Warming (2008) "Proposal of the Council on the Global Warming Issue, in Pursuit of Japan as a Low-Carbon Society", published in June 2008.

Den Elzen, Michel and Paul Lucas (2003) *FAIR 2.0 – A Decision-Support Tool to Assess the Environmental and Economic Consequences of Future Climate Regime*, Report 550015001/2003. Bilthoven: RIVM.

Dröge, Susanne (2009) *Tackling Leakage in a World of Unequal Carbon Prices*, Carbon Strategies Paper, 01 July 2009.

Egenhofer, Christian and Anton Georgiev (2009) *The Copenhagen Accord: A First Stab at Deciphering the Implications for the EU*, CEPS Commentary. 25 December 2009.

European Commission (2008) *20 20 by 2020 Europe's Climate Change Opportunity*, COM (2008) 30 final, Brussels, 23.1.2008.

Global Warming Prevention Headquarters (2008) *Implementation of Domestic Emissions Trading at a Trail Phase*. Available online at: http://www.env.go.jp/earth/ondanka/det/dim/trial/doc081021.pdf (accessed 31 July 2015).

Government of Japan (2007) *COP13, Gaiyo to Hyoka* [Summary and Evaluation]. Available online at: http://www.mofa.go.jp/mofaj/gaiko/kankyo/kiko/cop13_gh.html (accessed 31 July 2015).

Höhne, Niklas, Carolina Galleguillos, Kornelis Blok, Jochen Harnisch, and Dian Phylipsen (2003) *Evolution of Commitments under the UNFCCC: Involving Newly Industrialized Economies and Developing Countries*, Environmental Research of the Federal Ministry of the Environment, Nature Conservation and Nuclear Safety, Germany, Research Report 20141255.

Hook, Glenn D., Julie Gibson, Christopher W. Hughes and Hugo Dobson (2012) *Japan's International Relations: Politics, economics and security*. Abingdon: Routledge.

Industrial Structure Council (2003) *Perspectives and Actions to Construct a Future Sustainable Framework on Climate Change*, Interim report. Global Environmental Subcommittee Environmental Committee, Industrial Structure Council. Tokyo: METI.

Intergovernmental Panel on Climate Change (IPCC) (2007). *Climate Change 2007*. Cambridge: Cambridge University Press.

Kameyama, Yasuko (2007) "Process Matters: Building a Future Climate Regime with Multi-Processes," *Climate Policy*, 7(5), 429–443.

Kameyama, Yasuko (2008) "Chapter 2: Evolution of Debates over the "Beyond-2012" Climate Regime," in Yasuko Kameyama, Agus P. Sari, Moekti H. Soejachmoen, and Norichika Kanie eds., *Climate Change in Asia: Perspectives on the Future Climate Regime*. Tokyo: United Nations University Press, 18–30.

Keidanren (2008) "Toyako samitto ni okeru posuto kyoto giteisho no kokusai wakugumi kosho heno taio ni tsuite" [Comment regarding negotiation on the post-Kyoto multilateral framework to be discussed at the Toyako G8 Summit] a commentary posted on its website on 15 April 2008. Available online at: http://www.keidanren.or.jp/japanese/policy/2008/024.html (in Japanese) (accessed 2 July 2015).

Kiko Network (2009) "Chikyu ondanka no chuki mokuhyo ni tsuiteno iken" [Comments on the mid-term emission targets related to the global warming problem] a comment submitted to the Japanese government, 15 May 2009. Available online at: http://www.kikonet.org/info/press-release/2009–05–15/comment-on-2020-climate-target (in Japanese) (accessed 30 January 2016).

L'Aquila G8 summit (2008) G8 Leaders Declaration: Responsible Leadership for a Sustainable Future. Available online at: http://www.g8italia2009.it/static/G8_Allegato/G8_Declaration_08_07_09_final,0.pdf (accessed 10 August 2016).

Matsui, Tetsuya, Kiyoshi Takahashi, Nobuyuki Tanaka, Yasuaki Hijioka, Masahiro Horikawa, Tsutomu Yagihashi, and Hideo Harasawa (2009) "Evaluation of Habitat Sustainability and Vulnerability for Beech (Fagus crenata) Forests under 110 Hypothetical Climatic Change Scenarios in Japan," *Applied Vegetation Science*, 12, 328–339.

Meyer, Aubrey (2000) *Contraction & Convergence, The Global Solution to Climate Change*. Bristol: Green Books.

Ministry of the Environment (MOE) (2004) *2002nendo no onshitsu koka gasu haishutu-ryou ni tsuite* [Greenhouse gas emissions in the fiscal year 2002]. Available online at: http://www.env.go.jp/earth/ondanka/ghg/gaiyo.html (accessed 20 November 2015).

Ministry of the Environment (MOE) (2009) *Chuki Mokuhyo Kento Iinkai* [Committee on the Mid-term Emission Target]. Available online at: https://www.env.go.jp/earth/ondanka/mid-target/exam_prog.html (accessed 20 November 2015).

Ministry of Foreign Affairs (MOFA) (2008) *Official Website of Toyako G8 Summit.* Available on line at: http://www.mofa.go.jp/policy/economy/summit/2008/index.html (accessed 30 January 2016).

Nakamura, Hidenori, Mark Elder, and Hideyuki Mori (2011) "The Surprising Role of Local Governments in International Environmental Cooperation: The Case of Japanese Collaboration with Developing Countries," *Journal of Environment & Development*, 20(3), 219–250.

Nippon Keidanren (2008) *Keidanren Voluntary Action on the Environment* (first version published in 1997, periodically revised up to today). Available online at: http://www.keidanren.or.jp/english/policy/index07.html (accessed 30 January 2016).

Nordhaus, William (2001) *After Kyoto: Alternative Mechanisms to Control Global Warming*, paper prepared for a joint session of the American Economic Association and the Association of Environmental and Resource Economists, Atlanta, Georgia.

Ondanka Eikyo Yosoku Project (Global Warming Impact Estimation Project) (2009) *S-4 Ondanka Eikyo Sougou Yosoku Projekuto Saishu Hokokusho* [Project for Comprehensive Projection of Climate Change Impact: Final Report]. Available online at http://www.nies.go.jp/s4_impact/index.html (accessed 16 January 2016).

Ott, H. E., H. Winkler, B. Brouns, S. Kartha, M. Mace, S. Huq, Y. Kameyama, A.P. Sari, J. Pan, Y. Sokona, P.M. Bhandari, A. Kassenberg, E.L. La Rovere, and A. Rahman (2004) *South-North Dialogue on Equity in the Greenhouse*, a final report from South-North Dialogue project by Wuppertal Institute, Germany, and EDRC University of Cape Town, South Africa, supported by GTZ, Germany, May. 2004.

Prime Minister's Office (2007) "Invitation to Cool Earth 50" a speech made by then Prime Minister Shinzo Abe on 24 May 2007. Available online at: http://www.kantei.go.jp/jp/singi/ondanka/2007/0524inv/siryou2.pdf (accessed 16 January 2016).

Prime Minister's Office (2008a) Special Address by H.E. Mr. Yasuo Fukuda, Prime Minister of Japan On the Occasion of the Annual Meeting of the World Economic Forum (Cool Earth 50)" a speech made by then Prime Minister Yasuo Fukuda on 26 January 2008 at Davos, Switzerland. Available online at: http://www.mofa.go.jp/policy/economy/wef/2008/address-s.html (accessed 20 November 2015).

Prime Minister's Office (2008b) "Low Carbon Society and Japan (Fukuda Vision)" a speech made by then Prime Minister Yasuo Fukuda on 9 June 2008. Available online at: http://nettv.gov-online.go.jp/prg/prg1904.html (accessed 20 November 2015).

Prime Minister's Office (2009a) *Chikyu Ondanka Taisaku no Chuuki Mokuhyou ni Tsuite* [On Mid-term Target for Global Warming Measures] a document distributed in April 2009 prepared for Dialogue with the people. Available online at: http://www.kantei.go.jp/jp/singi/tikyuu/kaisai/dai09/09gijisidai.html (accessed 20 November 2015).

Prime Minister's Office (2009b) *Chikyu Ondanka Mondai ni Kansuru Kakuryo Iinkai* [Ministerial Committee on the Global Warming Problem]. Available online at: https://www.kantei.go.jp/jp/singi/t-ondanka/ (in Japanese) (accessed 30 January 2016).

Rudolph, Sven and Friedrich Schneider (2013) "Political Barriers of Implementing Carbon Markets in Japan: A Public Choice Analysis and the Empirical Evidence before and after the Fukushima Nuclear Disaster," *Environmental Economics and Policy Studies*, 15(2), 211–235.

Sampei, Yuki and Midori Aoyagi-Usui (2009) "Mass-Media Coverage, Its Influence on Public Awareness of Climate-Change Issues, and Implications for Japan's National Campaign to Reduce Greenhouse Gas Emissions," *Global Environmental Change*, 19, 203–212.

Sawa, Akihiro (2008) *A Sectoral Approach as an Option for a Post-Kyoto Framework*, The Harvard Project in International Climate Agreement Discussion Paper 08–23.

Schelling, Thomas (2002) "What Makes Greenhouse Sense? Time to Rethink the Kyoto Protocol," *Foreign Affairs*, 81(3), 2–9.

Schmidt, J., N. Helme, J. Lee, and M. Houdashelt (2006) *Sector-Based Approach to the Post-2012 Climate Change Policy Architecture*. Washington, DC: Center for Clean Air Policy.

Science Council of Japan (2009) *Chikyu Ondanka Mondai Kaiketsu no tameni: Chiken to Sesaku no Bunseki, Wareware no Torubeki Kodo no Sentakushi* [For the Solution of Global Warming Problem: Analysis on the Knowledge and Policies, Options for Actions to be Taken by Us], Report, 10 May 2009, Science Council of Japan.

Skjærseth, J.B. and J. Wettestad (2008) *EU Emissions Trading: Initiation, Decision-Making and Implementation*. Aldershot: Ashgate.

Stewart, Richard B. and Jonathan B. Wiener (2003) *Reconstructing Climate Policy: Beyond Kyoto*. Washington, DC: The AEI Press.

UNFCCC (2002) Delhi Ministerial Declaration on Climate Change and Sustainable Development, Decision1/CP.8, FCCC/CP/2002/7/Add.1.

UNFCCC (2004) Buenos Aires Programme of Work on Adaptation and Response Measures, Decision1/CP.10, FCCC/CP/2004/10/Add.1.

UNFCCC (2005a) Dialogue on Long-Term Cooperative Action to Address Climate Change by Enhancing Implementation of the Convention, Decision1/CP.11, FCCC/CP/2005/5/Add.1.

UNFCCC (2005b) Consideration of Commitments for Subsequent Periods for Parties Included in Annex I to the Convention under Article 3, Paragraph 9, of the Kyoto Protocol, Decision1/CMP.1, FCCC/KP/CMP/2005/8/Add.1.

UNFCCC (2007) Bali Action Plan, Decision 1/CP.13, FCCC/CP/2007/6/Add.1.

UNFCCC (2009) Copenhagen Accord, Decision 2/CP.15, FCCC/CP/2009/11/Add.1.

van Asselt, Harro (2014) *The Fragmentation of Global Climate Governance: Consequences and Management of Regime Interactions*. London: Edward Elgar Publishing.

Victor, David (2004) *The Collapse of the Kyoto Protocol and the Struggle to Slow Global Warming*. Princeton, NJ: Princeton University Press.

5 The Tohoku earthquake and reconsideration of Japan's energy policies (2011–2015)

Overview

Natsu kusa ya! Tsuwamono-domo ga Yume no ato
(Oh, summer weeds / all the soldiers / gone as a dream)

Basho Matsuo[1]

In the late seventeenth century, poet Basho Matsuo made a long journey in Tohoku, in northern Japan. In Hiraizumi, he recollected the story of a battle in the twelfth century in which a well-known leader of a fighting group, Yoshitsune Minamoto, was defeated after a gory battle. The scene must have been dreadful in the twelfth century, with dead bodies scattered on the ground, but after 500 years Basho saw only weeds growing peacefully in summertime. The battle appeared to be only a dream.

An earthquake and tsunami hit the Tohoku region on 11 March 2011, resulting in a serious nuclear accident at the Fukushima Daiichi nuclear power plant. Public opposition to the use of nuclear power plants became stronger and stronger, until a majority of the public opposed the use of nuclear power. All nuclear power plants in Japan were shut down and inspected following the accident. Because Japan's climate change mitigation policy at the time depended heavily on nuclear power, the roadmap toward low-carbon development had to be reconsidered. The climate change debate was more or less overtaken by the debate on nuclear energy policy. All of Fukushima Prefecture required fundamental reconstruction and restoration, at a time when the population of both Fukushima and Japan as a whole were decreasing. It was therefore difficult to estimate the appropriate amounts of effort and funding to be put into recovery efforts in Fukushima by the Japanese government.

At the international level, the Durban Platform was adopted in 2011. It initiated a new international negotiation to arrive at an agreement that would involve all parties. The Kyoto Protocol's second commitment period was officially confirmed in 2012. Together with Russia and New Zealand, Japan decided not to participate in the second commitment period but did not withdraw formally from the protocol.

Like Yoshitsune Minamoto more than 800 years ago, are Japan's nuclear power plants going to become little more than a myth? If so, how will Japan substantially reduce its greenhouse gas (GHG) emissions in the long term? Will various types of renewable energy eventually prevail, covering up the old myth of nuclear power like the summer weeds on the battlefield? Or will nuclear power regain credibility as a safe energy resource that contributes significantly to the reduction of carbon emissions? Fukushima and other prefectures in the Tohoku region are still waiting to receive financial support from the central government to recover from the disaster. Would it be possible to instill low-carbon development as a key concept in Tohoku's recovery to create renewable energy facilities in this area?

The Tohoku earthquake and reconsideration of energy policy

A magnitude 9 earthquake hit the northeastern region of Japan in the afternoon on 11 March 2011. It was followed by a ten meter tsunami that inundated 56,100 ha of land (JMA 2011; PMO 2012). Nearly 20,000 people were killed or lost during the disaster. The Fukushima Daiichi nuclear power plant on the seashore of Fukushima Prefecture was seriously damaged by the event, to the point that it reached meltdown. About 80,000 people living within a 20-kilometer radius of the plant had to be evacuated from their homes the next day, and some of those have not yet been able to return. Radioactive substances travelled over an extensive area, some reaching the Tokyo metropolitan area more than 200 kilometers away.

Before long, firm opposition to the use of nuclear power emerged throughout the country. Even before the accident, many people had questioned the cost effectiveness of nuclear power over other sources of energy, especially when the handling of radioactive wastes was taken into consideration (Oshima 2014). Before the accident, opponents of nuclear power had been unable to convince a majority of the public of this point of view. The majority accepted explanations from the central government asserting that nuclear power was the safest, cheapest, and cleanest source of energy.

In reality, nuclear power plants had been estimated to be the cheapest source of energy because the estimations excluded the costs of treating radioactive wastes and the costs that could be incurred in the case of serious accidents. Renewable energy sources such as solar and wind were considered too expensive, as compared with nuclear- and coal-powered plants. There was also little support from the government, even if the Japanese public had become more interested in renewable energy. As a result, the proportion of electricity generated by nuclear power plants reached about 26% in Japan before the earthquake. In addition, the government had planned to build more plants to supply 50% of all electricity, in part to fulfill Japan's 25% GHG emission reduction target.

The Japanese public's perception of nuclear power shifted after the accident. People demanded the immediate and total phase-out of nuclear power plants in Japan because, unlike in many other parts of the world, Japan was more likely to be hit by serious earthquakes. There were some industrial sectors, however, that still supported nuclear power, saying that such a serious earthquake was not likely to happen again in the near future, and also that the government could impose stricter standards to make nuclear power plants even safer than before.

After the earthquake and the accident at Fukushima Daiichi, energy policy became the upmost priority agenda in Japan. The then prime minister Naoto Kan said that Japan needed to "start from scratch" on its long-term energy policy (Reuters 2011). In May 2011, Kan ordered Kansai Electric Power to stop operation of the Hamaoka nuclear power plant, because Hamaoka was located in an earthquake-prone area. Other nuclear plants in Japan also ceased operations.

Nuclear and fossil fuels used to be the pillars of Japanese energy policy, but Prime Minister Kan was committed to putting more effort into the promotion of renewable energy sources, such as solar, wind, and biomass, as well as an increased focus on energy conservation. Promotion of renewable energy was one of the aims of the Democratic Party of Japan (DPJ) from the time of the previous prime minister, Yukio Hatoyama, but it started to gather more support from the public after the earthquake (*The Guardian* 2011).

A renewable energy bill was initially presented to the Diet in April 2011. The Renewable Portfolio Standard Law had been adopted in 2002 to enhance the use of renewable energy, but this law was not effective, because large-size electricity power companies were able to dominate the electricity market and could control the retail price of electricity by limiting the introduction of relatively expensive renewable energy to a minimum. The rate of renewables was kept low until the 2011 earthquake. Just before the 2011 disaster, hydropower accounted for 8.5% of the total electricity supply, and other renewable sources accounted for only 1.1%.

A *Special Law to Promote Renewable Energy* was enacted in August 2011, initially mainly to support solar energy, and a full-fledged feed-in tariff scheme was introduced in November 2011 to promote the installation of solar photovoltaics (PVs). The rules were revised in July 2012 to support other types of renewables. The new rules also obliged power companies to purchase all renewable energy that was generated. With this change, the share of renewable energy in the supply of electricity began to grow rapidly. The law also allowed power companies to place a surcharge on consumers' electricity bills to cover the cost required to purchase renewable energy. The introduction of the feed-in tariff was successful in rapidly increasing the installation of renewable energy – particularly solar PVs because of relatively higher amount of tariff compared to other types of renewables (JPY42 [around USD 0.4] per kWh for 20 years at the start of the scheme). Applications for new facilities approved by the Ministry of Economy, Trade, and Industry (METI) between July 2012 and December 2013 totaled 30 GW, with solar PVs accounting for 94% of the planned production (Kuramochi 2014). The scheme also boosted the installation of solar PVs on rooftops and in rural areas until power companies began to stop purchasing energy from these sources in fall 2014 owing to an oversupply of electricity from solar PVs.

A meeting of the National Stategy Bureau, a decision-making body established by the DPJ and consisting of the prime minister and other political and business leaders, was held in June 2011. The group adopted a decision to establish the Energy-Environment Council under the cabinet to develop a short-term strategy to fill the gap between demand and supply of electricity during the summer, when demand peaks, as well as to develop several options for energy supply in 2030. The Minister of the National Policy Unit was designated as chair, and the ministers of METI and the Ministry of Environment (MOE) were designated as cochairs.

The council first discussed ways to decrease electricity demand during summer by asking large-scale electricity consumers to minimize their activities. For example, railroad and subway companies in the metropolitan Tokyo area were asked to reduce the numbers of trains and subways operating during peak-demand hours. Caps were also set on electricity consumption by governmental institutions.

The Committee for Verification of Costs was set up under the council in October 2011 to discuss whether or not nuclear power was actually the least costly, and renewables were the most costly, sources of energy, as had been previously explained by METI officials and as had been generally believed by the Japanese people. The committee published its first report in December of the same year ((Energy – Environment Council 2011). It concluded that estimations of the costs of electricity generated by the respective energy resources were heavily affected by various assumptions, such as the consideration of damages in the case of serious incidents and of carbon prices. The committee revised the latest estimations from 2004, with higher costs for nuclear and fossil fuel incineration power plants and lower costs for renewable energy. Even so, nuclear and fossil fuel power plants were considered to be less expensive than renewables.

In the realm of climate change, the Hatoyama administration of the DPJ had aimed to agree on a new climate change bill called *the Basic Law Concerning Global Warming Measures*. The major difference between the previous *Law Concerning the Promotion of Measures to Cope With Global Warming* and the proposed new law was to set a legally binding emission reduction target of 25% by 2020 from the 1990 level, to set a long-term temperature increase target of 2 °C, and to confirm the introduction of major emission mitigation measures such as an emissions trading scheme and a carbon tax in the new law. The law continued to be supported by Prime Minister Naoto Kan, but the process was postponed by the earthquake, and Yoshihiko Noda, who succeeded Naoto Kan as prime minister in September 2011, was unable to push the legislation forward under pressure from the growing antinuclear movement. The proposal was finalized in the end of 2012 but resulted in no new legislation.

Durban Platform

Meanwhile, in the UNFCCC arena, many countries were seeing COP17 as the last opportunity for the UNFCCC parties to restart negotiations to adopt a new international framework for the future. In September 2011, Australia and Norway jointly submitted a proposal to adopt a decision at COP17 to initiate a new negotiating process aimed at adopting a new protocol in 2015 in which all countries would participate. The EU supported this idea as well, and it was dubbed the "Durban Road Map."

COP17 was held in Durban, South Africa, in December 2011. Three main agendas were intertwined. The first agenda was the "Durban Road Map." The second was the second commitment period of the Kyoto Protocol, and the third was the Green Climate Fund (GCF) established by the previous COP.

The EU was in the position that it could commit to the second commitment period of the Kyoto Protocol on the condition that the agreed-upon aim of the Durban Road Map was a new legally binding instrument, applicable to all parties. Because the EU had already invested in its own emissions trading scheme, a cap-and-trade type of regulatory framework outside the EU was considered preferable, leading to EU's support for continuation of the Kyoto Protocol. On the other hand, Japan, Canada, and Russia maintained the same position as the previous COP – they would not participate in the second commitment period of the Kyoto Protocol. Canada actually withdrew from the Kyoto Protocol just before COP17, disappointing many environmental nongovernmental organizations (NGOs). In other words, there was a great divide between the two camps of Annex I Kyoto parties on the issue of the second commitment period. On the other hand, both of these groups were in a coalition that asserted the importance of a new round of negotiations to establish a new instrument that would include all countries. Japan continued to assert its position to aim for an agreement that would include all major emitting countries.

The United States, which was not a party to the Kyoto Protocol, had little to say about the future of the protocol or of the second commitment period. It also had no objections to the idea of a new international framework. Nevertheless, it was concerned that, if the COP were to start another round of negotiations for a new international framework, the categorization of Annex I and non–Annex I countries should be totally phased out. The United States was of the opinion that the categorization had been agreed upon almost two decades ago in 1992, when there was a significant economic gap between developed and developing countries, but that the world had changed tremendously since that time. Some countries that were originally categorized as non–Annex I had successful emerging economies. These countries had even surpassed some of the Annex I countries in terms of gross domestic product (GDP) per capita. Thus, the categorization was viewed as outdated, at least by many developed countries.

Some emerging economies – namely, Brazil, China, India, and South Africa – had informally established a group called "BASIC" a few years earlier, but COP17 was the first opportunity for these countries to formally coordinate their positions. This support was helpful for the South African government, which was hosting the conference. The BASIC countries did not want to see a new protocol (at least not in the near future) and were opposed to the EU's idea of aiming for a protocol. Thus, the agreed-upon "Durban Platform" established the Ad Hoc Working Group on the Durban Platform for Enhanced Action (ADP), a subsidiary body with a mandate of helping parties develop a protocol or another legal instrument or an agreed-upon outcome with legal force under the convention applicable to all parties. This was to be completed no later than 2015 in order for it to be adopted at COP21 and for it to come into effect and be implemented by 2020 (UNFCCC

2011). By the same decision, the COP launched a work plan on enhancing mitigation ambition. It aimed to identify and explore options for a range of actions that could close the ambition gap with a view toward ensuring the highest possible mitigation efforts by all parties. The establishment of the ADP was basically welcomed by all countries, including Japan, that wanted to adopt a new legal instrument for the post-2020 period that could replace the Kyoto Protocol. On the other hand, many developing countries preferred the continuation of the Kyoto Protocol and believed that the new framework should not replace the protocol but only supplement it to cover actions by all countries.

In a way, the Durban Platform was similar to the Bali Action Plan, which aimed at an agreed outcome to be adopted in 2009. However, negotiators also did not want to see a repeat of the experiences of the Copenhagen meeting – they wanted something to actually be agreed upon this time.

Energy policy debate in 2012

Between December 2011 and June 2012, in Japan, the Energy-Environment Council made a thorough examination of the relationship between the use of nuclear power and CO_2 emission reductions. Three options were proposed by the MOE concerning the level of ambition for climate change mitigation policy. The most ambitious climate policy required rapid diffusion of the most energy-efficient instruments, using renewable energy, as well as changes to people's lifestyles and attitudes to save as much energy as possible. The least ambitious climate policy assumed the use of coal-powered electricity generation plants and no lifestyle changes. Of course, the former policy scenario required more initial investment than the latter.

In terms of energy policy, three options for the use of nuclear power were raised for possible emission reduction in 2030. The first option was a total phase-out of nuclear power as soon as possible, with 0% nuclear power in 2030. The second option was to reduce the share of nuclear power within the total electricity supply to 15% by 2030. The third was to reduce the share of nulcear power from the pre-earthquake period, but to maintain it at about 20%–25% in 2030. Once again, four modeling teams from the National Institute for Environmental Studies (NIES), Research Institute of Innovative Technology for the Earth (RITE), Osaka University, and Keio University were asked to make estimates of the cost of achieving each scenario. Once again, the results of the four models were different from each other, even when the same basic assumptions were used. In particular, calculations of GDP by NIES indicated a minor reduction in the case of additional policies as compared with the business-as-usual (BAU) case, whereas the RITE model indicated that additional climate policies would result in a major negative impact on GDP growth.

The government used various methods to collect people's views on the options. First, consultation meetings were held in 11 large cities in Japan, in which selected members of the public were allowed to present their own views in front of audiences. Second, public comments were collected via a website. A total of 89,124 comments were collected within one month. Third, a deliberative polling methodology was used to solicit more comments and preferences from about 7,000 everyday people on energy policy options (Energy – Environment Council 2012a. Finally, government officials individually visited 58 large businesses and labor organizations to explain the options and gather opinions of the audiences.

The cabient released its "Strategy for Innovative Energy and Environment" in September 2012 (Energy – Environment Council 2012c). The strategy aimed at reaching a zero-nuclear option in the future, but it also strongly emphasized further energy savings and the promotion of renewable energy, stating that these actions were indispensable to Japan's response to climate change. The proposed energy supply included greater use of more efficient liquified natural gas (LNG) and coal-fired power plants, which was expected to lead to growth in CO_2 emissions. Under the zero-nuclear option, it was considered almost impossible for Japan to meet its 25% emission reduction target (as agreed to under the Cancun Agreement) by 2020. The report said that, by 2030, Japan's CO_2 emission reduction would be about 20% of the 1990 level, at best. This target assumed that GDP growth was limited to 1% annually. If the GDP were to grow more than expected, such as at 2% per year, then the emission reduction potential would decrease to about 10% by 2030. Also, the report projected that, if the 20% reduction were to be achieved by 2030, the emission target for 2020 should be about a 5%–9% reduction from the 1990 level.

While the cabinet was busy trying to sort out the energy mix, MOE was making considerable progress in the long-lasting debate on a carbon tax, leading to the implementation of a tax in October 2012. The carbon tax and other environmental tax schemes had been discussed mainly within the domain of MOE since the mid-1990s, when some European countries began to introduce the tax. However, the idea was not well received by industry or most politicians. Prices of energy in Japan were already relatively high compared with those in other industrialized countries, largely because of the other taxes already in place. Carbon taxes had begun to be introduced in many countries in Europe, but opponents of the tax in Japan argued that the European carbon taxes were not effective in reducing CO_2 emissions, because large emitters in industrial sectors were exempt from the high tax rates. Japanese industries were also opposed to introduction of a tax because it would undermine the international competitiveness of Japanese industries.

Discussions on the carbon tax continued throughout the 2000s under committees established under MOE's authority. A proposal in 2004 was again strongly opposed by industry, and the committee was dismissed. Another proposal was made in 2005 after the first review of the *Kyoto Protocol Target Achievement Plan*, which concluded that additional measures were necessary to meet the 6% emissions reduction target. MOE proposed detailed rules for a carbon tax, at a rate of JPY2400/tC (around USD20), with the aim of introducing the tax in 2007. However, the proposal did not win public support because of increases in the price of gasoline on the international oil market and the resulting increases in the price of gasoline in Japan. Furthermore, the then prime minister Koizumi's position on "small government" worked as a disadvantage in the introduction of new taxes. It was only in 2009 that an excellent opportunity for the new carbon tax arose when a new ruling party, the DPJ, came into office and Prime Minister Hatoyama announced that Japan would aim at reducing GHG emissions by 25% from the 1990 level by 2020. Three economic instruments attracted the attention of the new ruling party: a carbon tax, an emissions trading scheme, and a feed-in tariff on renewable energy.

Industries argued that implementation of both emissions trading and a carbon tax would impose overlapping burdens on their activities. If they were to choose between the tax and emissions trading, Japanese industries preferred the tax over trading. This preference was based on the idea that only major emitting industrial sectors would be targeted for emissions trading, whereas the carbon tax could be implemented across all sectors, including households, at a relatively low rate. Although Keidanren and many other industry groups made appeals against the introduction of new taxes, Keizai Doyukai offered an alternative proposal, namely to terminate various existing taxes on energy and combine them into one consolidated environmental tax.

After the Fukushima Daiichi nuclear power plant accident and with stronger public support for the use of renewable energy, in addition to the Ministry of Finance being supportive of the introduction of new taxes to increase revenue, MOE incorporated a "global warming measures tax" in the tax plan for fiscal year (FY) 2012. The Japanese government then decided in late 2011 to introduce a carbon tax beginning in October 2012.

The carbon tax was applied to all types of fossil fuels, including coal, oil, and natural gas, in proportion to the amount of CO_2 they emitted. The tax rate was relatively low at JPY289/tCO_2 (a little less than USD3/tCO_2) in the initial implementation stage in 2012. The rate was to be raised gradually over the next 3.5 years. The tax revenue was ear-marked and was to be spent on the promotion of renewable energy and other means to improve energy efficiency. Given the relatively low tax rate, the small price increase of energy was generally thought to not change consumption patterns in the short term, but it was likely to affect consumers' choices of products in the long term. For example, people who used cars to commute would not shift to other means of transportation even if the price of gasoline increased somewhat, but they would probably buy a more energy-efficient car at their next purchase.

COP18 was held in Doha, Kuwait, in late 2012. Outcomes achieved through the conference included decisions on works of the ADP for the establishment of a new international framework; amendments to the Kyoto Protocol and the subsequent termination of the "Ad hoc Working Group on Further Commitments for Annex I Parties Under the Kyoto Protocol" (AWG-KP); decisions on long-term cooperation under the agreement and the subsequent termination of the Ad Hoc Working Group on Long-term Cooperative Action under the Convention (AWG-LCA); decisions on financing; and a COP decision on loss and damage associated with climate change impacts.

Under the AWG-LCA, the Cancun Agreement paved the way for a monitoring-reporting and verifying (MRV) process in 2020 that would ask countries to submit biannual reports to show their progress on mitigation and adaptation. In terms of the AWG-KP, the parties decided that the second commitment period would be between 2013 and 2020. Detailed rules such as those concerning the use of Clean Development Mechanism (CDM) emission reductions by countries not participating in the second commitment period were also determined. Countries not participating in the second commitment period, including Japan, were to be allowed to participate in CDM projects and acquire Certified Emission Reductions (CERs), but participation in joint implementation and emissions trading under the Kyoto Protocol during the second commitment period was allowed only for countries participating in the second commitment period.

In terms of ADP, basic arrangements for negotiations were set up toward reaching an agreement by 2015 on a new legal framework for beyond 2020. Decisions were made on the work plan for the next year onward, as well as on chair arrangements. Elements for a draft negotiating text were to be considered toward COP20 at the end of 2014, thus clarifying the steps for negotiations for the next year and beyond.

The Japanese government was most interested in highlighting its own carbon credit trading rule, namely the Joint Crediting Mechanism/Bilateral Offset Credit Mechanism (JCM/BOCM). Japan regarded this mechanism as complementary to CDM under the Kyoto Protocol. The architecture was quite similar to that of CDM, but with simplified procedures for approval. Japan also expected to incorporate the construction of nuclear power plants under the JCM/BOCM scheme (a type of project not approved by CDM). This was a controversial element of the JCM/BOCM even within Japan, because nuclear power plants had been criticized at home since the 2011 earthquake. Industries associated with nuclear power plant construction hoped to find new markets outside Japan. Japan started consultations on this mechanism with countries such as Indonesia, Vietnam, and Mongolia, with a view to begin operations by FY 2013. Japanese companies had already implemented more than 30 feasibility study projects in various sectors by the time of COP18. Japan's intention was to have the JCM/BOCM approved by the COP as a means of partial fulfillment of Japan's emission reduction target for 2020. This mechanism was related to the agenda known as "various approaches," which included discussions on allowing countries to take various measures to reduce emissions as long as they were credible.

Japan also emphasized its financial support for developing countries. Under the Copenhagen Accord, it was agreed that developed countries would contribute to "fast-start" finance, to provide USD30 billion for the three-year period from 2010 to 2012. Japan had pledged to provide a total of USD15 billion from both the public and private sectors over approximately three years, and it had achieved approximately USD17.4 billion in financing as of October 2012.

Organizational matters were discussed in relation to the Standing Committee of the GCF and endorsement of the host country of the GCF. Some developing countries pushed for a written description of concrete figures for mandatory support by developed countries in the coming several years, but this was opposed by the developed countries group and was deleted from the decision. On the other hand, all of the developed countries were urged to scale up climate finance from a wide variety of sources to reach a joint goal of mobilizing USD100 billion a year by 2020.

"Loss and damage associated with climate change impacts" became a hot agenda item, particularly for developing countries and small island states that were vulnerable to the adverse impacts of climate change. Despite many countries' recognition that some support was needed, the developed countries, including Japan, feared that the debate on loss and damage might develop into a demand for another new funding mechanism. The group eventually decided that a global mechanism to reduce damage in countries vulnerable to the adverse effects of climate change would be established at COP19.

Reconsideration of Japan's emission reduction target for 2020

The DPJ gradually lost the popularity and support that it had had at the time of its landslide victory in the election of summer 2009. This was perhaps partly a result of the earthquake and nuclear power plant accident, but also a result of its members' inexperience in decision-making as a ruling party. In December 2012, the Liberal Democratic Party (LDP) won the election of the Lower House Diet members. Shinzo Abe became prime qminister for the second time. As soon as Abe took his seat, he called for "Abenomics" or "Abeconomics," which were his strategies to prioritize stimulation of economic growth before anything else. On energy and climate policy, Abe stated he would "reconsider from zero-start," meaning that the GHG emission target for 2020, as well as the phasing-out of nuclear power plants, would be fully reconsidered. One of the key strategies Abe put in place was to reduce the yen's value against the US dollar in currency markets to increase exports of Japanese goods. Although a weaker yen against the dollar benefitted companies exporting goods, it also caused the prices of imported goods, including fossil fuel, to increase. Since it had phased out nuclear power plants, Japan had had to increase imports of fossil fuel, in particular natural gas (to minimize CO_2 emissions), and this resulted in increased energy costs. Already, by the time of the election campaign in late 2012, the LDP's position was to rehabilitate nuclear power to some extent. Industries were supportive of this position, considering that introduction of nuclear power would lead to a decrease in the cost of electricity. This situation further encouraged Japanese industries to increase their support for the use of nuclear power and for the LDP.

Nobuteru Ishihara was appointed Environment Minister by Abe at the time of Abe's inauguration as prime minister. He was the son of Shintaro Ishihara, former governor of the Tokyo Metropolitan Government (TMG), but that did not necessarily mean that Nobuteru was supportive of the emissions trading schemes that had originated in TMG. One of the first things he did was to reconsider Japan's emission reduction target for 2020. The previous 25% emission reduction target for 2020 had been set by former Prime Minister Hatoyama of the DPJ, and the LDP considered the target to be too ambitious and unrealistic, even before the earthquake and nuclear accident. Now that the future of nuclear power had become uncertain, the LDP was eager to revisit the GHG emission target for 2020. Even during DPJ's reign, the report "Strategy for Innovative Energy and Environment" in September 2012 (Energy – Environment Council 2012c) suggested that a phase-out of nuclear power plants would lead to Japan not being able to fulfill its 25% emission reduction target; the report also noted inconsistencies between the energy strategy and the climate change emission reduction target. The DPJ did not have an opportunity to respond to the report, but the LDP utilized the same report as background data to insist on the need to use nuclear power plants and to revise the 2020 emission reduction target.

In March 2013, the cabinet adopted a decision to revise the *Law Concerning the Promotion of Measures to Cope With Global Warming* to incorporate a climate change mitigation plan for the years beyond 2013. This revision was not a progressive one, because it prescribed neither emission targets for 2020 nor the long-term target of an 80% emission reduction target by 2050, which had already been accepted by the cabinet as well as the Global Warming Prevention Headquarters. A debate between MOE and METI began on whether or not to determine a revised emission reduction target for 2020. MOE considered it important to announce a new target for 2020, because even if the target were to be notably lower than the 25% Hatoyama target it would still be better than no target. With no clearly quantified target, MOE would not be able to justify its assertions of the need to implement additional emission reduction measures in Japan. METI, on the other hand, argued that, because the level of use of nuclear power was not likely to be confirmed in the near future, Japan should abandon its emission reduction target for 2020 and start discussions on a new target for 2030, after Japan's energy policy had been determined.

The election for the Upper House Diet members was conducted in July 2014. Each political party published a "manifesto," a document consisting of commitments each party promised to implement if it won the election. Policies on climate change and nuclear power were mentioned in the manifestos of all of the major political parties. The Japan Communist Party and Social Democratic Party asserted the need for ambitious emission reduction targets and efforts to expand renewable energy. The LDP showed little interest in climate change mitigation actions and emphasized that its priorities were economic growth and the creation of new jobs. The LDP, still under the Abe administration, won the support of a large majority of the people. The election results reflected a low point in public interest in global climate change in Japan.

Working Group I of the IPCC concluded its fifth assessment report in September 2013. In the report, it stated that "warming of the climate system is unequivocal, and since the 1950s, many of the observed changes are unprecedented over decades to millennia. The atmosphere and ocean have warmed, the amounts of snow and ice have diminished, sea level has risen, and the concentrations of GHGs have increased" (IPCC 2013). This message was noted by the government in Japan, and it decided to set a revised emission reduction target that would be announced at COP19. However, there was no time to go through a detailed decision-making procedure, such as requesting modeling groups to conduct another round of calculations or calling for a meeting of experts to discuss the issue in detail.

To put the decision-making in context, as a result of the switch from nuclear power to fossil fuels, carbon dioxide (CO_2) emissions in Japan were steadily on the rise. Total GHG emissions increased by 2.7% from 2011 to 2012 and by an additional 1.3% from 2012 to 2013. In 2013, emission levels were 10.6% greater than those in 1990 (MOE 2014a). Cabinet members associated with the climate change debate gathered in early October, and they internally discussed a new target that would be set, but 2005 would be the base year, not 1990. Because GHG emissions in Japan grew about 7% between 1990 and 2005, the same reduction target (in terms of percent) for 2020 would be quite different if the base year were changed from 1990 to 2005.

COP19 was held in Warsaw from 11 to 23 November 2013. Environment Minister Ishihara and officials from relevant ministries attended the meeting. Japan brought two key messages to the Conference: a changed reduction target and a diplomatic strategy. Ishihara proposed a new emission target of a 3.8% reduction by 2020 from the 2005 level (MOE 2013). He said that this new target (which appeared to be less ambitious) was at least a target at this point. The ultimate effects of the allocation of nuclear power had not yet been fully accounted for because of uncertainty. He proposed that a firm target would eventually be set on the basis of a further review of Japan's energy policy and energy mix, particularly on the use of nuclear power. Ishihara emphasized that the target was ambitious and that it required a 20% improvement in energy efficiency at all levels of the Japanese economy. A target for the land use, land-use changes, and forestry sector was proposed to ensure that the amount of CO_2 sequestered by forest management for the period between FY2013 and FY2020 would be an average of 3.5% of the total GHG emissions in FY 2005. Japan promoted the establishment and implementation of the JCM, and its target assumed that some of the emission reduction was to be gained from its use. The JCM scheme had previously been called JCM/BOCM, but the government decided to delete the term "BOCM" because the idea of "off-setting" was already included in the definition of "crediting" (MOE 2015a).

The diplomatic message was to propose a proactive strategy to counter global warming – it was titled "ACE: Actions for Cool Earth." In an attempt to reach the goal to reduce emissions by 50% at the global level and by 80% in the developed world by 2050, Japan announced that it would promote further technological innovation. It would promote emission reductions in developing countries by diffusing advanced low-carbon technologies through the JCM. Furthermore, in the area of mitigation of, and adaptation to, climate change, Japan would provide support to developing countries totaling JPY1.6 trillion (about USD16 billion) during the three-year period from 2013 to 2015, making use of Official Development Assistance, Other Official Flows, and private flows.

These messages were meant to convey the idea that the Japanese government was doing its best to contribute to climate change mitigation, but it was criticized by some as inefficient and inadequate. Particularly, the new emission reduction target of "3.8% from 2005" was equal to a 3% increase from the 1990 level. This meant that Japan intended to increase its GHG emissions from the end of the first commitment period of the Kyoto Protocol. Although the new target did not include the use of any nuclear power, the Japanese government did not explain how the emission target would be revised if the nuclear power were accepted by the Japanese people in the future. In addition, the timing of announcement was unfortunate. ADP Workstream 2 had been discussing how to get countries to enhance their Cancun emission targets for 2020. The total amount of pledges for 2020 had been proven to be insufficient to reach the 2 °C goal, and more mitigation actions were therefore needed. Because Japan lowered the level of its pledge, it was criticized by some countries and environmental NGOs because they said it needed to raise the level of its targets, not lower them.

Regarding COP19 as a whole, ADP Workstream 1 discussed general features of each country's commitment that would be scribed in the agreement. Some countries preferred wording such as "commitment" to indicate that actions taken by countries would be legally binding. Others preferred wording such as "actions" so that all countries – including developing countries – would participate in the agreement. Another point of conflict was the date of submission of information related to commitments to the UNFCCC secretariat. Those who were concerned that the internally determined emission targets would not be sufficient to mitigate climate change thought that some type of ex-ante consultation process would be needed before confirmation of the levels. The consultation process could assess whether or not emission reduction commitments submitted by countries were sufficient, and, if not, a country could reconsider its target. The commitments needed to be submitted quite early if this type of consultation process were to be conducted before COP21.

The final decision stated:

> [We] invite all Parties to initiate or intensify domestic preparations for their intended nationally determined contributions, without prejudice to the legal nature of the contributions, in the context of adopting a protocol, another legal instrument or an agreed outcome with legal force under the Convention as set out in its Article 2 and to communicate them well in advance of COP21 (by the first quarter of 2015 by those Parties ready to do so) in a manner that facilitates the clarity, transparency and understanding of the intended contributions, without prejudice to the legal nature of the contributions.
>
> (UNFCCC 2013)

On the issue of finance, the developing countries' group insisted that, in order to be certain that the developed countries jointly reached the USD100 billion funding target by 2020, COP should set a midterm goal of USD70 billion by 2016. The developed countries were strongly opposed to this idea, and the two groups ended up agreeing on a reporting requirement every two years from 2014 to 2020 on the level of financial resources. A package of rules on Reducing Emissions from Deforestation and Degradation in developing countries (REDD+) was reached at

COP19. The package included a procedure to maintain transparency and ways to facilitate financial support for REDD+ related projects.

Loss and damage were becoming increasingly important, particularly for vulnerable developing countries. These countries asserted a need for the establishment of a new funding mechanism or insurance to cover loss and damage due to climate change, but developed countries held the position that such an additional institution or mechanism was not required, at least in the short term. The COP adopted a decision to establish the new "Warsaw International Mechanism for Loss and Damage associated with Climate Change Impacts (Loss and Damage Mechanism)," to address loss and damage associated with impacts of climate change, including extreme events and slow onset events, in developing countries that are particularly vulnerable to the adverse effects of climate change. The mechanism would deal with the enhancement of knowledge and understanding of comprehensive risk-management approaches to address loss and damage associated with the adverse effects of climate change, including slow-onset impacts, and the strengthening of dialogue, coordination, coherence, and synergies among relevant stakeholders.

Reconstruction of Japan's energy policy for the post-2020 period

As the Abe Administration entered its second year, the Japanese government slowly began to discuss its future energy policy. On 11 April 2014, the cabinet decided to approve the new Basic Energy Plan as the basis for Japan's new energy policy, with consideration of the dramatic changes in energy environments inside and outside Japan, including those caused by the Great East Japan Earthquake and the subsequent events at TEPCO's Fukushima Daiichi Nuclear Power Station (METI 2014). At the beginning, the plan described Japan's vulnerable energy situation because of the country's need to import most of its energy resources. The plan also indicated concerns about the increasing cost of energy resulting from the phase-out of nuclear power, and about the increase in imports of natural gas and crude oil. The plan noted that 83% of crude oil and 30% of natural gas imported by Japan were from the Middle East, and it also noted the political instability of that region. The plan presented the basic principles of Japan's energy policy as "3E+S" – that is, "energy security," "improving economic efficiency," "environment suitability," and "safety," and indicated the basic direction of the energy policy.

The plan went on to assess each source of energy. Renewable energy was thought to be important, but further consideration was needed because of its high cost. Nuclear power was described as "an important base-load power source as a low carbon and quasi-domestic energy source, contributing to stability of [the] energy supply-demand structure, on the major premise of ensuring . . . its safety." Coal was acknowledged as the cheapest source of energy for which a stable supply was available without much geopolitical risk. The plan expected coal could be further utilized by introducing highly efficient technologies, such as integrated coal gasification combined cycle technology. The plan also sought to export high-efficiency coal power plants.

While debates on climate change were being reframed as debates on nuclear power, developments in terms of other aspects of emission mitigation at the domestic level were making little progress after the earthquake. For instance, the LDP took office with a landslide victory in the December 2012 election, when it opposed renewable energy, which was deemed too expensive, claiming that the high cost of electricity affected energy-intensive industries and low-income households. Power companies started to build new small-scale coal-fired plants to reduce costs, leading to considerable growth in Japan's CO_2 emissions. These companies also temporarily halted their purchase of electricity from solar power firms in 2013 and 2014, on the grounds that with their limited transmission capacity they could not sustain a rapid increase in power supply from such unstable sources. Power companies had been reluctant to invest in increasing capacity to transfer electricity from renewables, because purchasing electricity from various generation endpoints would decrease their own sales of electricity. This problem had been anticipated since the early days of implementation of the feed-in tariff, and supporters of renewable energy had long been asserting the need to increase the capacity to transfer electricity on Japan's electricity grids.

In other sectors of the Japanese economy, many new climate mitigation policies were also making little progress. In the industrial sector, total CO_2 emissions decreased by 6% from 2005 to 2013 (Figure 5.1). Most of this reduction, however, was attributed to the economic crisis of 2008 and confusion in the aftermath of the 2011 earthquake. This indicates that the reduction was achieved by a decrease in production of goods, rather than by improvements in energy efficiency. Moreover, the level of improvements in energy efficiency differed from one industry to another. Energy efficiencies in the materials industries, such as iron and steel, cement, and paper and pulp, were mostly stable over this period, meaning that CO_2 emissions from these industries were affected mainly by production output.

In the residential and commercial sectors, the Japanese government had revised the *Law Concerning the Rational Use of Energy* several times, so that industries had to use their production processes to equip their products with the best available technologies. A variety of subsidies were granted for energy-efficient products, including electrical home appliances, buildings, houses, and cars. Many Japanese home appliances made extensive improvements in energy efficiencies because of Japan's "top-runner" program, a policy in which manufacturers had to meet the latest standards of the most advanced devices. The average refrigerator and air-conditioner sold in 2013, for example, consumed 67% and 21% less electricity, respectively, than those sold in 2003. Among the measures implemented was a type of subsidy known as "eco-points" for the purchase of appliances with increased energy efficiency.

In addition, the government of Japan called for "electricity saving" in 2011 and 2012 to adjust to the insufficient power supply that had occurred in the aftermath of the earthquake. Given people's generally high awareness of the issue and increasing

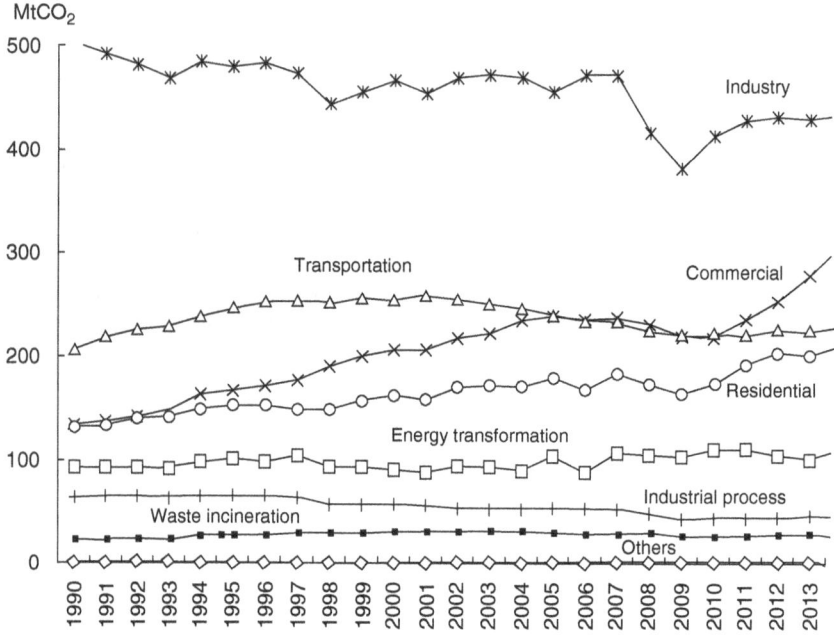

Figure 5.1 Japan's CO$_2$ emissions by sector
Source: MOE (2015b)

electricity prices, electricity-saving home appliances – such as refrigerators, air conditioners, TV sets, and LED lamps – sold well. After people had replaced these types of appliances, however, their overall awareness of importance of energy saving faded. Furthermore, as the installation of coal power plants increased, power shortage ceased to exist, leading to further shrink in people's care for energy saving. Improvements in the insulation of buildings and houses had begun in the early 2000s, and energy management systems were utilized to reduce unnecessary use of energy. A subsidy known as "housing eco-points" was also introduced for those who built houses by using better-insulated windows and construction materials.

Despite these efforts and achievements in energy efficiency in the residential and commercial sectors, CO$_2$ emissions from these sectors increased by 11.9% and 16.7%, respectively, from 2005 to 2013. The growth in the residential sector was mainly due to continuing increase in the number of household in Japan. Number of people per household has been decreasing, but each household requires minimum outline of facilities such as refrigerators and TV sets. The growth in the commercial sector was attributed mainly to an increase in floor space for offices, leading to a demand for more air conditioning and lighting.

In the transportation sector, CO$_2$ emissions grew throughout the 1990s, but they peaked in about 2000–2001 and thereafter began to decrease gradually, leading to a reduction of 6.3% from 2005 to 2013. Since the first hybrid car from Toyota debuted in 1995, hybrid and compact cars have been subsidized and have experienced healthy sales. Electric cars (considered the cleanest cars in terms of CO_2) have also attracted much attention; this was especially the case before the earthquake, when the price of electricity was expected to decrease due to increase in supply from nuclear power plants. Since the earthquake, however, more people have selected hybrid cars. Some electric cars have batteries that can be charged by solar PV cells on the cars' roof-tops, thus promoting the use of renewable energy. Such usage is, however, not yet widespread.

In early 2014, the Japanese government finalized GHG emission data for the year 2012, and proved that Japan successfully achieved the 6% reduction target set by the Kyoto Protocol. The average of GHG emissions between 2008 and 2012 was 1.4% above the 1990 level, whereas sequestration by LULUCF and use of Kyoto Mechanisms accounted for 3.9% and 5.9%, respectively.

COP20 in Lima

As Japan kept relatively quiet in climate-related debates, primarily because of the public's sensitivity on the subject of nuclear energy, other countries gradually began to move toward arriving at a new multilateral agreement at COP21, to be held in Paris in December 2015.

On 12 November 2014, President Barack Obama of the United States and President Xi Jinping of China jointly made an announcement on climate change mitigation strategies, reaffirming the importance of strengthening bilateral cooperation on climate change and their commitment to work together to adopt an agreement at COP21. President Obama announced a new target of the United States to cut its overall GHG emissions to 26%–28% below 2005 levels by 2025. President Xi Jinping also announced China's target to reach peak CO_2 emissions by about 2030, with the intention to try to do even earlier, and to increase the non–fossil-fuel share of all energy to about 20% by 2030. This initiative of the world's two largest GHG emitters came as a surprise to many Japanese stakeholders, because they had long considered these two countries as lagging behind in the realm of climate change negotiations. In reality, many Japanese people were not that well-informed of events outside of Japan, as the country continued to struggle to get over the nightmare of the March 11 nuclear power accident.

COP20 took place in Lima, Peru, in December 2014. In terms of the ADP process, participants in the Lima Conference faced three themes. The first was on plausible elements of the Paris agreement for the post-2020 period. The second was on the schedule of the negotiation process between COP20 and COP21. At COP 19, it was determined that each country was to submit its intended nationally determined contribution (INDC) in early 2015, preferably before 31 March 2015.

However, there had been no agreement on what kind of information should be included in the INDC reports or what should be done with the submitted INDCs before COP21. The developed countries thought that the information should be focused mainly on mitigation actions, such as the underlying assumptions for the estimation of future GHG emissions, and on the core mitigation policies that were being used to reach the emission targets. On the other hand, developing countries thought that information related to adaptation and the means of implementation – particularly those related to financial contributions – should also be included. In addition, major countries such as the United States, other Umbrella Group countries (Australia, Canada, New Zealand, Kazakhstan, Norway, the Russian Federation, Ukraine, and the United States), and many developing countries did not want any ex-ante consultation process to assess the adequacy of each country's INDC, whereas the EU and many small island states asserted that some type of ex-ante consultation process was needed because the submitted levels in the INDCs possibly could be close to the BAU emission trajectory, reduction effort insufficient to meet the desired goals.

The third theme was on ADP Workstream 2, which aimed at further deepening countries' levels of ambition for their 2020 targets. UNEP's Emission Gap Report (UNEP 2014) had indicated that the total emission level targets of each country under the Cancun Agreement for 2020 would not lead to the long-term global temperature-increase target of 2 °C. A series of Technical Expert Meetings (TEMs) was held during COP19 and COP20, and they were generally considered to be successful. However, countries had different ideas as to what should be done after the TEMs were over. Developing countries argued that developed countries should be responsible for making further contributions, not only in mitigation but also in supporting developing countries financially and technically in the pre-2020 period. The developed countries thought that mitigation ought to be the main scope of what was indicated in the Durban Platform, so that other items such as adaptation and finance was out of the Workstream 2 mandate.

On the final day, the "Lima Call for Climate Action" was adopted (UNFCCC 2014). It called for countries to submit INDCs that included quantifiable information on the reference point; timeframes and/or periods for implementation, scope, and coverage; planning processes, assumptions, and methodological approaches (including those for estimating and accounting for GHGs); reasons why the party considered its INDC to be fair and ambitious, in light of its national circumstances; and ways in which the INDC helped to achieve the objective of the convention as set out in Article 2. The same decision also requested that the UNFCCC secretariat publish the INDC as communicated on the UNFCCC website, and prepare by 1 November 2015 a synthesis report on the aggregate effect of the INDCs communicated by the parties by 1 October 2015. Although some countries asserted the importance of conducting an ex-ante consultation on the INDCs to discuss the levels of ambition of the contributions, this concept was not included in the decision. The decision also attached an annex concerning elements for a draft negotiating text, which was 39 pages long. It included all of the parties' submissions, sorted out by items, with brackets as available options. The discussion did not result in any meaningful decisions relevant to Workstream 2, apart from continuing the TEMs process.

During the Lima conference, Japan's position was mostly consistent with those of other members of the Umbrella Group. The group members sought an agreement that would include the participation of all countries and would not divide countries into groups, such as Annex I and non–Annex I (i.e., developed and developing countries). They considered that such a division may have been justified in the 1980s but that countries were more diverse today. This group was happy with the idea that emission reduction targets would not be legally binding in nature, but they also proposed various ways in which the agreement could motivate countries to achieve their respective emission reduction targets. Japan thought that it was important to submit progress reports and update them regularly. Others, including Russia, were reluctant to establish a burdensome review process. Finally, the group members had a consolidated view on financial assistance to developing countries – that no numerical targets should be set on financial mobilization and that no additional financial scheme should be constructed under the "loss and damage" agenda.

In Japan, the general perception was that it would be difficult and expensive to reduce GHG emissions at home. At the same time, many stakeholders believed that exporting Japanese products and technologies to developing countries to help reduce emissions in those countries was a better strategy for use by Japan to respond to the climate change problem. A speech by Environment Minister Yoshio Mochizuki at COP20 emphasized Japan's financial and technical contributions to developing countries for the reduction of GHG emissions and achievement of low-carbon development (MOE 2014b). Specifically, Japan had established the unique JCM approach. There were 12 JCM partner countries (Bangladesh, Cambodia, Costa Rica, Ethiopia, Indonesia, Kenya, Laos, Maldives, Mexico, Mongolia, Palau, and Vietnam) at the time of COP20, and Minister Mochizuki stated that this type of crediting scheme would help host countries to reduce GHG emissions, save energy, and enhance economic development. The Japanese government hoped that this mechanism would be accepted at the UNFCCC level in both the pre- and the post-2020 periods so that Japan could count on credit acquisitions to achieve its emission reduction targets.

Determination of emission reduction targets for 2030

A formal discussion process on Japan's INDC started in October 2014 with the establishment of a Joint Council Meeting consisting of two key committees, the Global Environmental Committee under the Central Environmental Council and the Technology and Environment for Industries Committee under the Industry Structure Council (MOE 2014c). The MOE was responsible for the former committee, whereas the METI was responsible for the latter. The Joint Council Meeting of the Central Environment Council and the Industrial Structure Council was cochaired by Professor Naoto Asano (Fukuoka University) and Professor Kenji Yamaji (Research Institute of Innovative Technology for the Earth). The Joint Council Meeting consisted of 20 members, including the cochairs, with backgrounds in academia, research institutions, the business sector, and labor unions.

Some members of the Global Environmental Committee emphasized the importance of achieving the Copenhagen Accord's long-term target of a 2 °C maximum increase in global temperature, as well as of Japan's role in international negotiations on equity perspectives. Others from the Technology and Environment for Industries Committee, however, argued that any global emission reduction targets could be achieved only with significant contributions from large emitters, such as

China and the United States, and that Japan, which was then responsible for about 3% of global emissions, could not make any notable changes at the global level simply by reducing domestic emissions. These members were fully aware at the time that they did not have the ability to determine Japan's INDC singlehandedly, because other governmental committees and working groups under METI and the Agency for Natural Resources and Energy (ANRE) were also discussing the future energy mix.

Generally speaking, there are two types of approaches to determining levels of emission reduction targets at the national level. The first is a top-down approach, in which a cap for GHG emissions is set at a global level – for example, the level that would be required to limit global GHG emissions to achieve the 2 °C target. Then, some kind of equity criterion is applied to distribute emissions to each country. This approach is effective in determining the degree of emission reduction that needs to be achieved to mitigate climate change. The second is a bottom-up approach, in which emission reduction potentials are gathered from each subsector – even from the product level. The reductions are then summed to ascertain how much emission reduction would be achievable. This approach is effective in determining the foreseeable potential emission reduction. In most cases, the target levels set by the first approach are much lower (more stringent) than those set by the latter approach. The 25% emission reduction by 2020, set by the Hatoyama Administration in 2009, was arrived at by using the first approach. The Prime Minister himself set that target, and, afterward, government officials and researchers tried to come up with the energy mix needed in Japan in 2020 to reach the target.

The Abe administration clearly took the latter approach in discussions of Japan's emission reduction target for 2030, similar to what most of his predecessors in the LDP had done previously. METI was particularly eager to revise the decision-making approach, and it wanted to determine an energy plan first before setting a GHG emission reduction target. Therefore, METI took the lead in establishing various committees and subcommittees to initiate discussions on future energy policy. A subcommittee on energy efficiency, energy saving, and renewable energy; another subcommittee on electricity and gas; a nuclear power committee; a committee to determine the price of renewable energy; and a working group on power grids all conducted parallel meetings on this topic. Until these committees, subcommittees, and working groups came to a conclusion, the Joint Council Meeting on INDC could not start meaningful discussions on the reduction target in Japan's INDC.

Despite the fact that GHG emissions from the electricity-generating sector accounted for less than half of Japan's total GHG emissions, the country's climate change policy debate continued to focus on the controversial issue of the use of nuclear power plants. Before the earthquake, nuclear power constituted about 25%–30% of Japan's total electricity supply (World Nuclear Association 2015). Since the earthquake, natural gas and coal had been used to provide the portion of the supply that had been previously generated by nuclear power.

The government began speeding up consideration on Japan's INDC after the Lima Call for Climate Action was adopted at COP20 in December 2014. Japan, however, was not ready to submit its INDC by the end of March because, as noted previously, any discussion related to the INDC in Japan involved extensive debate on future energy policy, which was not yet settled. The Japanese government aimed to determine its INDC by the end of May so that Prime Minister Abe would be able to announce the target numbers at the Group of 7 (G7) Summit in Germany on 7 and 8 June.

Prime Minister Abe had repeatedly pledged that, although his administration might not completely do away with nuclear power, he sought to reduce Japan's dependency on nuclear power to the extent possible, largely to maintain his popularity. Some LDP members held different views. They viewed economic growth as the top priority, and they wanted to see nuclear power return to its previous position as a "base-load" energy resource, meaning that the resource was considered a core of Japan's energy supply. The ambitiousness of the INDC was used as a justification to rehabilitate nuclear power plants. A schedule for meetings for the various committees and councils was set up so that the energy mix for 2030 would be determined by the end of April, leaving very little time for the Joint Council Meeting to discuss adequate target levels to meet the requirements of the INDC by the end of May.

The Third UN World Conference on Disaster Risk Reduction was held in Sendai in Miyagi Prefecture in March 2015. Originally, this conference was established in light of an earthquake that had occurred in Hyogo Prefecture in 1995. Since then, Japan had taken the lead in holding conferences on natural disasters. However, Japan also considered this forum to be a part of its contribution in the area of adaptation to climate change. Prime Minister Abe pledged USD4 billion to support implementation of the "Sendai Cooperation Initiative for Disaster Risk Reduction" over the next four years. The French foreign minister, Laurent Fabius, who had been appointed president of COP21, was also present at the conference. He made an appeal for the creation of a worldwide early warning system for climate disasters, because he said that 70% of disasters were now linked to climate change – double the number of 20 years ago. Such discussions were expected to stimulate Japan's attention to COP21, but they seemed to have had little effect on the energy mix debate.

A proposal on the energy mix within the electricity sector in 2030 was officially made public by ANRE on 27 April 2015. The proposal suggested that, in 2030, 20%–22% of electricity should be supplied by nuclear, 22%–24% by renewables, 26% by coal, 27% by LNG, and 3% by oil. This proposal required that about 36 or 37 of the 43 existing but currently shut-down nuclear power plants be utilized to meet the goal. In FY 2012, the share of LNG-fired thermal power had reached 42.5%, whereas that of coal was 27.6%. In the proposed 2030 energy mix, the government sought to reduce the share of natural gas, retain the same amount of coal-fired power generation (despite coal's greater environmental impact), and put nuclear power plants back online to cover the gap created by the reduction in natural gas–powered electricity output.

Three days later, on 30 April, the government disclosed its proposal for Japan's INDC, which was a 26% reduction of GHG emissions by 2030 from the 2013 level. This was equal to a 25.4% reduction from 2005 and an 18.0% reduction from 1990. Although the choice of base year did not affect the absolute amount of emissions in 2030, it conveyed a message. By choosing 2013 as the base year, the Japanese government insisted that Japan's INDC was more ambitious than that of the United States and the EU. As can be seen in Table 5.1, the EU's reduction rate would look the most ambitious if 1990 were chosen as the base year, whereas the US proposal would look the most ambitious if 2005 were used as the base year. The Japanese government argued that each country was choosing a base year that would put its own proposal in the best light. This argument could, however, be challenged by the EU, which continued to reduce its GHG emissions throughout the period from 1990 to 2015. Reportedly, Prime Minister Abe gave three instructions to be followed when determining the energy mix and the target level in the INDC (Nippon Keizai Shimbun 2015). The first was to lower the price of electricity from current levels. One of the reasons why the government maintained a large share of power generation from coal-fired plants was to reduce generation costs. The second instruction was to increase the share of renewable energy so that the share would be larger than that of nuclear power, with the final values of 22%–24% and 20%–22% chosen for renewables and nuclear, respectively. The third instruction was to set Japan's INDC at a level that was competitive with those of the United States and the EU. Choosing 2013 as the base year highlighted the fact that Japan's target was competitive with those of the other countries.

Table 5.1 Intended Nationally Determined Contributions (INDCs) of major economies

Base year	Rate of change from the base year (%)		
	1990	2005	2013
Japan: reducing GHG emissions by 26% of 2013 levels by 2030	−18.0	−25.4	−26.0
United States: reducing GHG emissions by 26%–28% of 2005 levels by 2025	−14 to −16	−26 to −28	−18 to −21
EU28: reducing GHG emissions by 40% of 1990 levels by 2030	−40	-35	−24
Russia: reducing GHG emissions by 25%–30% of 1990 levels by 2030	−25 to −30	+10 to +18 (increase)	—
China: reaching peak CO_2 emissions by 2030	+277 to +316 (increase)	+82 to +101 (increase)	—

Note: the figures in the table were estimated by using the RITE model.

Source: Yamaji (2015)

The Japanese government finalized its INDC on 1 June and invited comments from the public between 3 June and 2 July.

After MOE and METI had finalized their discussions on INDCs, MOE began to pressure companies that were planning to build new coal power plants in Japan to reconsider their plans. As part of the deregulation of the electricity industry, from April 2016 onward the electricity market was to become a free market. Many companies had already started planning to build new plants that generated electricity in the cheapest way – coal power plants. MOE was concerned that Japan would surpass its target share for coal within total electricity generation in 2030 if all of the planned coal power plants were built and operated between now and 2030. Although METI basically agreed with MOE's position, it remained silent on the issue. On 12 June 2015, Environment Minister Mochizuki requested that METI Minister Yoichi Miyazawa stop the planned construction of a new coal power plant in Ube by Osaka Gas, J-Power, and Ube Kosan. The Environment Minister claimed that building new coal power plants would harm efforts to meet the INDC target for 2030. The plant in Ube was only one of 10 new coal power plants that were on the waiting list for approval.

MOE, however, had no authority to directly regulate the activities of power companies. The only tool it had was an environmental assessment procedure, but even this process had been used in such a way that power companies were being officially accredited by the government to build new power plants. MOE sought to strengthen the environmental assessment procedure to limit the number of new coal power plants, but this issue was resolved in February 2016 in a way favorable for the power companies. The construction of new coal-fired power plants was also criticized by environmental NGOs. The NGO E3G, which is based in Europe, published a scorecard for G7 member countries by benchmarking coal phase-out actions. The report indicated that Germany and Japan were the two countries that still relied heavily on coal, and the situation in Japan was worse because Japan was still planning to build new plants (E3G 2015).

On 16 July, METI officially announced Japan's prospects for long-term energy supply and demand. With this publication of Japan's energy prospects, debate on energy policy in Japan was essentially concluded. The content was consistent with the discussions and conclusions of the various committees and councils during spring. The Japanese government also submitted Japan's INDC to the UNFCCC secretariat the next day, on 17 July (Government of Japan 2015). Although the government took one month to collect public comments, the final INDC was exactly the same as the original one put forth in late May. The INDC stated:

> Japan's INDC towards post-2020 GHG emission reductions is at the level of a reduction of 26.0% by FY 2030 compared to FY 2013 (25.4% reduction compared to FY 2005) (approximately 1.042 billion t-CO_2 eq. as 2030 emissions), ensuring consistency with its energy mix, set as a feasible reduction target by bottom-up calculation with concrete policies, measures and individual technologies taking into adequate consideration, *inter alia*, technological and cost constraints, and set based on the amount of domestic emission reductions and removals assumed to be obtained.

In August, the Sendai nuclear power plant in Kyushu in the southern region of Japan was switched back on. This was the first restart since nuclear power plants in Japan had been shut down in the wake of the Fukushima Daiichi disaster four years earlier. To begin operations again, the Sendai plant faced increased scrutiny from regulators and the courts. The Japanese government overhauled its nuclear safety policy, reviewed its atomic energy infrastructure, and created the new Nuclear Regulation Authority. The reopening of the Sendai nuclear power plant was received with mixed feelings by the public. On the one hand, many people still resisted the thought of any use of nuclear power, but other people were convinced that some nuclear power was needed to better utilize facilities that had already been built and lower the cost of electricity generation. In addition, it was also expected to help the world's third-largest economy to lower its carbon emissions.

Negotiation processes at the international level and COP21

The negotiation process gradually started to pick up its pace to arrive at an agreement by the end of December 2015. ADP2–8 was held in February 2015 in Geneva. This was the last chance for countries to develop a consolidated negotiating text for COP21, because it was supposed to be ready at least six months before COP21. Countries generally worked in a constructive manner, and the meeting was able to conclude with a 90-page negotiating text known as the Geneva Negotiating Text (GNT), which became the starting point of further negotiations.

During the meeting, Japan held positions consistent with those of COP20. Japan considered that mitigation actions should be more focused than other types of actions such as adaptation, finance, and technology transfer, whereas the developing country group emphasized that all types of commitments by developed countries should be balanced. For example, they believed that financial and technical support for developing countries' mitigation and adaptation actions should be an integral part of developed countries' commitments.

Japan had not officially submitted its INDC to the UNFCCC secretariat at the time of ADP2–9, which was held in June 2015 in Bonn, but Prime Minister Abe had unofficially mentioned its planned figures at the Elmau G7 Summit, which was also held in Germany. Countries did not publicly comment on the Japanese INDC, but environmental NGOs criticized it as inadequate to meet the goal of reaching the 2 °C long-term temperature increase; they also criticized it for the use of 2013 as the base year and for the proposed high rate of use of coal power plants (CAT 2015). The Elmau G7 Summit finalized a declaration stating that "we emphasize that deep cuts in global GHG emission are required with a de-carbonization of the global economy over the course of this century," and "we support sharing with all Parties to the UNFCCC the upper end of the latest IPCC recommendation of a 40 to 70% reduction by 2050 compared to 2010 recognizing that this challenge can only be met by a global response." The media reported that Prime Minister Abe was reluctant to agree to these statements, especially because he had been interested in maintaining coal power plants in Japan as well as in building new relatively more efficient coal power plants in other developing Asian countries (Stefanini 2015).

ADP2–9 was held in conjunction with UNFCCC's 42th Subsidiary Bodies' meeting (SB42) in June 2015 in Bonn. The main objective of the meeting was to walk through the GNT and see if there was any room for streamlining it – in other words, to see if any overlapping paragraphs could be shortened or moved to other sections of the text. The negotiating text was categorized into 12 sections, as shown in Table 5.2.

During the SB42 meeting, Japan was also one of the countries assessed under the Multilateral Assessment process set by the Cancun Agreement. Countries commented on Japan's first biannual report. Comments included the following: "We are disappointed with Japan's changing the 2020 target from '-25% from 1990' to a '3.8% reduction from 2005.' When is Japan intending to confirm the latest target?"; "Is Japan considering any new mechanisms not to repeat backsliding the emission targets?"; "We understand Japan intends to use JCM credits to meet its 2020 target. How?"; "There are already CDM credits which could be used to meet the 2020 target. Why does Japan want to use JCM credits in addition? How many JCM credits are likely to be used by 2020?"; and "How many nuclear power plants are likely to be restored by 2020, and how would that affect the GHG emission trend in Japan?" The Japanese government was unable to provide a specific timetable for the future use of nuclear power; it also had difficulties responding to other comments with concrete figures and to questions of whether other mitigation policies were also related to the level of use of nuclear power.

While Japan was still struggling to get out of aftermath of nuclear power plant accident, the world seemed to be becoming more interested in the climate change problem. For example, Pope Francis issued an encyclical on climate change entitled *"Laudato Si,"* or "Praise Be to You," in June 2015. The document, nearly 200-pages long, emphasized that climate change was an urgent problem and called for a new dialogue about how we were shaping the future of our planet. The discussion especially focused on the future of the world's poorest, who were

Table 5.2 The Geneva Negotiating Text (GNT) by section

A. Preamble
B. Definitions
C. General matters, objectives
D. Mitigation
E. Adaptation, loss, and damage
F. Finance
G. Technology development and transfer
H. Capacity-building
I. Transparency of action and support
J. Commitment periods, and process
K. Implementation and compliance
L. Procedural matters
Annex

most affected by a changing climate. This was the first time that the head of a large religious group had publicized his position on climate change.

In July, after the ADP2–9 meeting, the ADP cochairs, Daniel Reifsnyder and Ahmed Djoghlaf, developed a "nonpaper" – a kind of unofficial document that did not require countries' consensus as an official document but that helped countries to engage in discussions without burdensome procedures. Countries again gathered in Bonn later in the summer to make further progress at ADP2–10. Countries worked in spin-off groups that were established mainly according to article numbers. Countries found themselves in constructive discussions, and they asked the cochairs on the last day of the meeting to create another version of the nonpaper that reflected discussions held during the ADP2–10 meeting.

Summer vacation was over, and people started preparing for more intense negotiation toward COP21. In conjunction with the UN event, "United Nations Sustainable Development Summit 2015," held on 25–27 September, political leaders also gathered in New York to discuss the climate change problem. Before the Summit, President Xi Jinping of China visited President Obama in Washington, D.C., to reaffirm their shared conviction that each country has a critical role to play in addressing climate change. In a joint statement, President Xi confirmed that China would introduce an emissions trading scheme at the national level in 2017 and that it would cover key industrial sectors such as iron and steel, power generation, chemicals, building materials, paper-making, and nonferrous metals.

One and a half months after the ADP2–10 meeting, countries met for the last time before COP21, on 19–23 October in Bonn for ADP2–11. The cochairs drafted a nonpaper on 5 October and posted it on the organization's website. The draft was a simple and concise document, with only nine pages for the draft agreement, 11 pages for draft decisions, and nine pages for draft decisions on Workstream 2. Many countries expressed the view that the cochairs' draft deleted too many elements that had appeared in the nonpaper of the ADP2–10 meeting and that it did not reflect the discussions held there. Therefore, countries worked to reinsert some paragraphs and options that they considered indispensable. The paper's length increased to 34 pages, including COP decisions and decisions related to Workstream 2. The following four days were allocated to starting negotiations, but the participants ended up only revising the text to streamline it, avoid duplication, and clarify the options available in each article. Although this was considered to be a necessary part of the process, the speed of the work was not fast enough to finalize a text with fewer options and brackets. The only remaining time to do so would be at COP21.

On 30 October 2015, the UNFCCC secretariat published a synthesis report on the agregate effect of the INDCs communicated by 147 parties (UNFCCC 2015a). This represented 75% of the parties and 86% of global emissions in 2010. The aggregate effect on the INDCs until 2030 was characterized as follows: "The implementation of the communicated INDCs is estimated to result in an aggregate global emission level of 55.2 Gt CO_2 eq in 2025 and 56.7 Gt CO_2 eq in 2030." Implementation of the INDCs, therefore, would be effective in lowering aggregate global emission levels as compared with pre-INDC trajectories and slowing down the speed of growth in emissions in the 2010–2030 period relative to that in the

1990–2010 period. This was good news. The level of aggregate INDCs was, however, not sufficient to reach the long-term goal of a 2 °C temperature increase. It was clear that the current planned trajectory would not be sufficient to minimize the adverse effects of climate change.

In mid-November, an OECD working group agreed that its member countries would restrict export financing to build coal power plants overseas, with limited exceptions. The World Bank, as well as public financial institutions in the EU and the United States, had previously announced that it would, in principle, cease lending to finance such projects, and private-sector lenders in these countries were beginning to follow suit. These movements were called "divestments", a notion which had not been shared by most of Japanese stakeholders until then. Under the new OECD agreement, which takes effect in 2017, financing would still be allowed for the most advanced "ultra-supercritical" plants and for some other plants in the very poorest countries. Japan insisted that coal was the cheapest source of energy and thus was wanted by many developing countries. The European countries and the United States, however, considered coal power plants to be harmful, not only with respect to climate change but also in terms of local air pollution, in addition to the health hazards posed by coal mines. The OECD therefore allowed only those plants with the highest level of efficiency to compromise with Japanese assertion. Although this debate was held only at the OECD level, the decision sent a message to climate change negotiators under the UNFCCC that use of coal was no longer supported.

Just before the OECD decision and only a few weeks before COP21, a group of terrorists opened fire at multiple locations in Paris, killing more than 120 people. Despite the security concerns posed by COP21 and the many heads of state that were scheduled to attend the opening ceremony, the French government and the UNFCCC secretariat decided to convene COP21 as scheduled.

On 28 November, the Japanese cabinet adopted Japan's adaptation plan. This was the first time that the Japanese government had officially addressed adaptation measures. It took MOE almost a year to put the plan together. It was initially scheduled to be adopted in summer, but this deadline was postponed because of conflict between MOE, which wanted to see adaptation measures placed in firm legislation, and MITI, which wanted to limit the adaptation plan to a cabinet decision and limit the scope of legislation to mitigation only. Because it was quite obvious that the adaptation plan would work as an excellent opportunity to increase the budget of relevant agents, other ministries such as the Ministry of Agriculture, Fisheries, and Forests (MAFF) and the Ministry of Land, Infrastructure, Transport and Tourism (MLIT) asserted their dominance over adaptation plans in their respective sections.

COP21 opened on 30 November 2015. Unlike at previous COPs, the organizers of this conference invited heads of state to speak on the opening day. This was partially a result of learning from the experience of Copenhagen in 2009 (COP15), where heads of state had been invited on the final day, and the conference ultimately had not been successful. In Paris, heads of state made opening statements emphasizing the significance of coming to an agreement at this COP, but they were not fully involved in the actual drafting of the agreement. President Obama emphasized that the United States had been investing in renewable energy for the past seven years, helping create areas in the United States where clean (i.e., renewable) power sources were actually cheaper than conventional fossil fuel power plants. He emphasized that the country was succeeding in decoupling emission reductions from economic growth. In his speech, Prime Minister Abe emphasized Japan's financial support for developing contries. He said that Japan would provide, by 2020, a total of approximately JPY1.3 trillon (approximately USD10 billion) of public and private climate finance to developing countries.

The negotiations on the post-2020 framework made some progress over the course of next two weeks. In the second week, an informal gathering called "Comité de Paris" was organized to finalize the text of the agreement. The three most contentious remaining issues were diffentiation between realtively wealthy and poor countries, financing for both the pre- and post-2020 period, and raising the levels of ambition of the target goals. In terms of the first issue, Japan clearly was in favor of "self-differentiation," in which the commitments of the developed countries would not be decisively segregated from those of the developing countries, but the developed countries would voluntarily commit to more ambitious commitments than those of the developing countries. However, this position was criticized by the developing countries' group, headed by India, Malaysia, and other countries. On the second issue, Japan and other developed countries wanted to see more countries contribute to financial support. Even within the non–Annex I group, there were countries that had made rapid economic growth since the 1990s and had become relatively wealthy. Many developing countries thought that the developed countries should remain the only parties contributed to financing. Furthermore, for the post-2020 period the developing countries wanted to see concrete numerical financial goals, which the developed countries would not accept. For the third issue, Japan took the position that the INDC should be determined by each country and should not be subject to international consultation before finalization. Japan supported regular reporting and reviewing (e.g., every five years). Japan was not enthusiastic about setting a long-term temperature goal of a 1.5 °C increase, because it thought that the goal was unrealistic and unachievable.

Japan also had other concerns and preferences. One was on the entry-into-force issue. Japan's priority was the participation by all major emitters, and it wanted to ensure this by underscoring the importance of setting a "double trigger," in terms of both number of countries participating and share of total global emissions. Another concern was the use of carbon markets. Japan was committed to increasing the number of JCM projects in the pre-2020 period to partially fulfill the Cancun target for 2020, and it preferred the use of this type of crediting mechanism in the post-2020 period. Compared with the atmosphere during the negotiations towards the Kyoto Protocol, the positions of METI, MOE, and the Ministry of Foreign Affairs (MOFA) were not much divergent in Paris. Japan's emission reduction targets for 2030 had been determined in July 2015, and its adaptation plan had been adopted in November, so all of the major issues had been resolved within the Japanese delegation by the time of COP21.

During the two weeks of negotiation, the EU and some small island states formed an informal group known as the "high ambition coalition." This group aimed at inserting the 1.5 °C goal and the concept of decarbonization of the economy into the agreement text. The number of countries joining this coalition grew daily, with the United States and Brazil joining in the second week. Intensive negotiations continued throughout the second week, and the agreement was finally reached in the evening of 12 December 2015.

The text, called the Paris Agreement (UNFCCC 2015b), was not called a protocol, but it was nonetheless a legal document as an international treaty. In Article 2, the long-term goal was set as keeping the "global average temperature to well below 2 °C above pre-industrial levels and to pursue efforts to limit the temperature increase to 1.5 °C above pre-industrial levels." Commitments on mitigation were stipulated in Article 4, which stated, "Each party shall prepare, communicate and maintain successive nationally determined contributions that it intends to achieve. Parties shall pursue domestic mitigation measures, with the aim of achieving the objectives of such contributions." This meant that no clear distinction between "developed" and "developing" countries was mentioned. Concerns of the developing countries were reflected in other paragraphs that said, "Developed country Parties should continue taking the lead by undertaking economy-wide absolute emission reduction targets," and "[s]upport shall be provided to developing country Parties for the implementation of this Article, in accordance with Articles 9, 10 and 11, recognizing that enhanced support for developing country Parties will allow for higher ambition in their actions."

As for finance, the developed countries' assertion was reflected in the second paragraph that said "other parties are encouraged to provide or continue to provide such support voluntarily." The developing countries' assertion to set a numerical target for the post-2020 financial contribution was not mentioned in the agreement text but was inserted in a COP decision, stating, "developed countries intend to continue their existing collective mobilization goal through 2025 in the context of meaningful mitigation actions and transparency on implementation; prior to 2025 the Conference of the Parties serving as the meeting of the Parties to the Paris Agreement shall set a new collective quantified goal from a floor of USD100 billion per year, taking into account the needs and priorities of developing countries." Japan's position was supported for entry into force. Article 21 stated, "This agreement shall enter into force on the 30th day after the date on which at least 55 Parties to the Convention accounting in total for at least an estimated 55% of the total global GHG emissions have deposited their instruments of ratification, acceptance, approval or accession."

The Paris Agreement was basically welcomed by all countries, including Japan. Japan got most of what it wanted: participation by all countries, the dissolution of any clear division between developed and developing countries, the nonlegally binding nature of nationally determined emission reduction targets without any ex-ante consultation process, and a periodic review of progress for all countries. Japan was also pleased to see the inclusion of an article that allowed the

JCM to be utilized in the post-2020 period to acquire emission credits from other countries. A clear indication of the need of achieving a balance between emissions and removals of GHGs in the second half of the twenty-first century was welcomed mostly by the MOE. These benefits, however, were obtained without the exertion of strong leadship by Japan. Rather, most of them were also preferable for most of the other developed countries.

Eventually after the Paris Agreement, countries will participate in a process to review total amount of their INDCs to evaluate whether the INDCs were sufficient to mitigate climate change to a level that would not harm Earth's ecosystem. This type of global stocktaking was a first step forward, but there were no clear guidelines as to what would be done if countries' mitigation efforts proved to be insufficient.

Involvement of political leaders

Prime Minister Naoto Kan faced several diplomatic challenges during his relatively brief administration. In September 2010, a Chinese fishing boat intentionally collided with a boat from the Japanese coast guard near the Senkaku Islands, which had been claimed as national territory by both Japan and China. The captain of the boat was arrested but was soon released and sent back to China. This incident sparked controversy in Japan. Those opposed to Kan thought that Japan should not have released the captain so quickly. Prime Minister Kan met with US president Obama to highlight the strong relationship between the United States and Japan. Kan's favorable position on locating the US military base in Futenma (in Okinawa) helped him to reassure the US president of Japan's desire for a good relationship with the United States. Kan also had a bilateral meeting with Hu Jintao, then president of China, in November 2010 during the Asia-Pacific Economic Cooperation meeting in Yokohama, but the meeting was not productive, with both sides insisting on their sovereignty over the Senkaku Islands. The aggravating relationship between China and Japan hampered Japan to promote any kind of partnership with China concerning climate change mitigation activities.

The earthquake of March 11 was a turning point for Kan. After that, he had little time to spare on foreign affairs. He had been supportive of renewable energy even before the earthquake, but his support for renewables became even greater after the earthquake. He had also been somewhat supportive of the use of nuclear power, but this changed completely after the Fukushima Daiichi accident. Two months after the accident, Kan decided to close the Hamaoka nuclear power plant because it was in the Tokai area, which is at high risk of experiencing a serious earthquake in the near future. He went on to instruct all nuclear power plants to terminate operations for inspection in 2012. Introduction of the feed-in tariff in 2012 was Kan's final contribution to support for renewable energy before his resignation in September 2011.

Ryu Matsumoto was appointed Environment Minister in September 2010 and served through June 2011. He put in a great deal of effort at the COP10 of the Convention on Biological Diversity, held in Nagoya in October 2010. The conference successfully adopted the Nagoya Protocol on Access and

Benefit-Sharing. With this experience, Minister Matsumoto had more interest in biodiversity-related events than those on climate change. After the earthquake in 2011, he was also appointed minister of reconstruction, responsible for reconstruction after the disaster in Fukushima Prefecture. The next environment minister, Satsuki Eda, held office for only three months and had little opportunity to exert any influence.

Yoshihiko Noda took office as prime minister in September 2011. Dealing with the aftermath of the earthquake continued to be a serious burden on his administration. Apart from energy policy, his foremost priority was to increase the consumption tax rate. Japan's expenditure had been increasing because of expanding demand for social welfare services (especially for the rapidly aging population), and reconstruction efforts in Fukushima Prefecture imposed an additional financial burden in a time of serious budgetary constraints. In August 2012, the Diet accepted his proposal to raise the consumption tax rate from 5% to 8% in April 2014 and from 8% to 10% in October 2015. This was contemporaneous with the introduction of a carbon tax in October 2012, which meant that the carbon tax was supported by the prime minster as a way to complement the consumption tax in reducing government debt. On the whole, however, Noda focused his political attention almost entirely on domestic matters, so that (compared with his predecessor) he paid relatively little attention to foreign policy matters, including climate change.

Goshi Hosoda was appointed as the environment minister in September 2011 and stayed in the position through October 2012. By then, MOE was responsible for the regulation of nuclear and radioactive waste management, and the newly appointed ministers were more interested in nuclear-related issues than in climate change and other traditional environmental problems. Goshi Hosoda was one of a group of politicians who were eager to tackle problems related to the aftermath of the nuclear power plant accident and reconstruct a safer society. The next Environment Minister, Hiroyuki Nagahama, was in the position for only two months.

Prime Minister Abe took office for the second time in December 2012 in a landslide victory by the LDP in the Lower House Diet member election. Because it was his second time in office, he was even more firmly committed to making changes in many disputed areas. His central concern was generating economic growth in Japan, and one of his aims was to weaken the exchange rate of the Japanese yen against the US dollar. A relatively weak yen would result in increased exports of Japanese products, which was a primary goal of Abe. At the same time, a weaker yen against foreign currencies was a disadvantage for importers of natural resources, including fossil fuels. Of the fossil fuels, natural gas emitted the least CO_2, so imports of natural gas increased during these years, but natural gas was also more expensive than coal and oil per unit of energy obtained. Thus, the payments for natural gas grew as both demand increased and the currency exchange rate shifted in the direction of a weaker yen. Rehabilitation of nuclear power was therefore considered indispensable for two reasons: to reduce spending on imports of expensive fossil fuels and to reduce CO_2 emissions as a part of climate change mitigation.

Prime Minister Abe was also passionate about making progress in the area of national security. His ideal solution was to revise Article 9 of the Japanese constitution that renounced war and stated, "Aspiring sincerely to an international peace based on justice and order, the Japanese people forever renounce war as a sovereign right of the nation and the threat or use of force as means of settling international disputes. In order to accomplish the aim of the preceding paragraph, land, sea, and air forces, as well as other war potential, will never be maintained. The right of belligerency of the state will not be recognized." However, revision of the constitution was considered very unrealistic under the current circumstances. Therefore, in 2015, Prime Minister Abe pushed for amendments to legislation regarding decision-making procedures when the cabinet wanted to send Japanese troops overseas, without making any amendments to the Constitution. As a huge crowd of protesters gathered outside the Diet building, the Upper House passed the amendments in September 2015. The shift in the direction of preparing stronger military forces was partially in line with the growing global concern over the increased influence of Islamic State. For example, two Japanese men were captured by the Islamic State members in early 2015, but Japan was unable save them and they were soon killed.

Abe basically followed the path of his predecessor in raising the consumption tax rate from 8% to 10%, but he did postpone the scheduled introduction from October 2015 to April 2016 to gain taxpayers' support. In addition, the Komei Party, a member of his political alliance, wanted the Abe administration to push for tax relief on food products. Any exemption was deemed by the Ministry of Finance (MOF) and the LDP Tax Commission to be unacceptable, but Abe finalized the exemption in a cabinet decision in December 2015 accepting some areas of exemptions.

Abe's preference for a lower consumption tax rate with exceptions was inevitable, because his priority was on stronger economic growth. This stance was also consistent with his position on climate change policy, in which he was more interested in mobilizing public and private money to invest in the installation of Japanese energy-efficient technology in developing countries, so that Japan would gain economic benefit from the transactions. Abe particularly favored the use of highly efficient coal power plants. Under the Abe administration, Japan stood out among the developed nations in continuing to extend financial aid to developing countries seeking to build coal-fired power plants. The Abe administration, in its quest to boost infrastructure exports as a key to the nation's economic growth, pushed overseas sales of Japan's coal-fired power plant technology, with the government-funded Japan Bank for International Cooperation extending large loans for projects in countries such as Indonesia, Thailand, and the Philippines. The government's point of view was that replacing the old power plants in other countries with the latest, highly efficient coal-fired plants would contribute to reducing global emissions of GHGs. The government has not since backed off from this assertion even after a round of criticisms made at multilateral meetings regarding the use of coal.

Nobuteru Ishihara, the environment minister appointed under the Abe administration, stayed in office for two years, between December 2012 and

September 2014. This relatively long period in power enabled him to better understand the status quo of the climate change problem and the position of Japan in the negotiations. Both Prime Minister Abe and Ishihara were willing to change the 25% GHG emission reduction target for 2020, set by former prime minister Hatoyama of the DPJ. Because all the nuclear power plants were under inspection at the time of their inauguration and there was no energy plan in place for the future, they felt there was no way to confirm Japan's ability to reach the emission target by 2020. Environment Minister Ishihara announced that the 25% target was no more applicable and that he intended to announce a revised target before COP19. The revised target that he eventually announced at COP19 was not ambitious, but he did note that another revision could be made when the nuclear power policy was finally established.

Yoshio Mochizuki was appointed environment minister in September 2014 and held office through September 2015. The term was too short to influence government decisions on the 2030 GHG emission targets, but he took the lead in requesting that electric power companies reconsider their plans to build new coal-fired power plants. The next environment minister, Tamayo Marukawa, attended the Paris conference (COP21). Marukawa was relatively young and had a background in broadcast journalism. Although she had little experience in the environmental field, her English-speaking skills were of great use in helping to finalize the agreement.

Involvement of nonstate actors

Local initiatives

The TMG continued to take the lead in local initiatives in Japan on climate change mitigation policies. After Shintaro Ishihara's long reign as governor, Naoki Inose held the post for a year from December 2012 to December 2013. Inose's background was as a writer and journalist, not a politician. However, he reportedly had vested interests in highway road construction dating back to 2001 during the Koizumi administration, and he slowly began to take part in politics. His most well-known accomplishment during his administration was the International Olympic Committee's decision to have Tokyo host the 2020 Olympic Games. Inose was popular among Tokyo's citizens, but he was forced to resign when he was suspected of receiving bribes from a medical group.

The next governor, Yoichi Masuzoe, came into office in February 2014. He had experience as the minister of welfare at the national level, so his major interests were in Japan's aging society and in child care, not climate change. It was thus not the governor himself but other officials in TMG's Environment Bureau who continued its work on climate change policy. In March 2014, TMG established an additional target: reducing energy consumption 20% by 2020 from the 2000 level. The TMG had previously set a GHG emission target of a "25% reduction by 2020 from 2000 levels." Although energy consumption had been steadily decreasing, CO_2 emissions were rising because of the increased use of fossil fuels in the electricity sector.

The TMG is currently scheduled to continue its cap-and-trade scheme in a second phase (2015–2019). In addition, in January 2015, TMG announced its energy consumption goal of a 30% decrease from the 2000 level by 2030. The TMG set the target for energy consumption, not for GHG emissions, because there was so much uncertainty regarding the future energy supply. Any ambitious CO_2 emission reduction target would be much more easily achievable if nuclear power plants were fully back in operation by 2030. Energy consumption in the TMG area has decreased steadily since 2001, whereas regional GDP has grown, indicating the area's success in decoupling energy use and economic growth. Tokyo will host the Olympic Games in 2020, and a committee has already been established to discuss ways to minimize CO_2 emissions related to the preparation and hosting of the games. In addition, TMG drafted its environment plan in October 2015 and proposed a new GHG emission reduction target of "30% from the year 2000." This was equal to a 28% reduction from 2013, which was perceived as more ambitious than the national INDC. TMG expected that more hydrogen energy will be supplied by 2030 and that emissions from the commercial sector, which were still increasing, would continue to be capped by the emissions trading scheme.

Many other local governments also started internal discussions to set emission goals for 2030. Cities such as Fukui, Kitakyushu, Kyoto, and Yokohama are well known for their high levels of environmental awareness and their ambitions to set stringent targets. However, many of these government authorities are pressured by local industries and commercial coalitions, making it difficult to set goals that are more ambitious than the national goal. Thus, a majority of local governments are likely to settle for setting local targets equal to the national target.

For most local governments in Japan, climate change policy has meant mitigation policies. They have not been obliged to set any adaptation measures, mainly because MOE has never asked them to do so. Even if some local government officials in charge of environmental issues have attempted to discuss adaptation, they no doubt would have been opposing voices from other departments stating that any extreme weather patterns may not be caused by climate change and that environmental departments have no authority over infrastructure or city planning. It was only after MOE started drafting its adaptation plan in late 2015 that local governments also started to include adaptation strategies in their own climate change planning. Japan, as a whole, was late in arriving to the area of adaptation measures.

Industry

Japanese industry was worried about the high cost of electricity after the earthquake. Many companies wanted to see nuclear power plants back online, and they also wanted to see more coal power plants built rather than increased use of renewable energy, because coal was viewed as the cheapest source of energy. In April 2015, Keidanren published a commentary on Japan's energy mix, stating that Japan should have a sufficient "base-load" electricity supply; this meant that at least 60% of electricity should come from either nuclear, fossil fuels, hydro-power, or geothermal. It recommended that the distribution should be 60% fosssil fuels (including 27% coal), 25% nuclear, 10% hydro and geothermal, and 5% from other renewables. The report noted that, according to a modeling exercise by RITE, an increase of 5% in renewables in the electricity supply would cost the Japanese economy JPY600–1100 billion (about USD5–10 billion; Keidan-ren 2015a). Keidanren issued another commentary in September 2015 about the upcoming Paris meeting (Keidanren 2015b). In it, the group asserted the need for a new framework to be applicable to all, and that the INDCs should not be legally binding. It praised the Japanese goverment for stating that Japan would not par-ticipate in the second commitment period of the Kyoto Protocol on the opening day of COP16 in 2010. Keidanren claimed that Japan's opening statement was the turning point in UNFCCC negotiations to arrive at the Durban Platform, which called for a new framework applicable to all parties. The comment went on to say the bottom-up approach rather than top-down approach should be applied to assess the country's effort to mitigate climate change. Although the theme of the commentary was the international framework, as part of its environmental assess-ment it also included a paragraph commenting on MOE's interventions against building new coal-fired power plants. Keidanren insisted that environmental assessment was a communication tool for sharing views among industries, gov-ernments, and people living near project sites, and that it should not be used to regulate CO_2 emissions.

"Denjiren" is a Japanese abbreviation for the Federation of Electric Power Companies of Japan. This group exerts political power to influence Japan's power supply, and it affected the government's decision-making on Japan's energy mix for 2030, as well as for the GHG emission target for the same year. It emphasized a "balance between environment and economy" as the baseline and emphasized the importance of the wise use of nuclear power under strict safety standards, as well as improved efficiency in coal-fired power plants.

Meanwhile, some business groups began to shift away from the Keidanren position. For example, the Japan Climate Leaders' Partnership (Japan-CLP) was formed in 2009 to tackle climate change from a business perspective. The group published a commentary in May 2015 (Japan-CLP 2015) stating that it would be desirable for Japan's midterm reduction target to be more ambitious to better align with the long-term 80% reduction target by 2050 that had already been approved by the Japanese cabinet. It also stated that "ambitious climate action is not merely a cost, but is a promising investment to maintain the basis of people's livelihoods, and strengthen the competitiveness of Japan, a country with limited resources." The group was allied with a European-based coalition of private companies known as We Mean Business, which also aimed at low-carbon development (We Mean Business 2015). Another example was Toyota, which announced its 2050 target to reduce CO_2 emissions from newly sold cars by 90% from today's (year 2015) level. Companies such as Toyota, which has many stakeholders outside Japan, need to maintain their good images and to take the preferences of overseas consumers into account.

NGOs

Kiko Network and WWF Japan continued to be the leading environmental NGOs in submitting comments to the government on national policy-making and decisions, publicizing briefings for the general public and the mass media, and conducting symposia and other events related to climate change to raise public awareness. Kiko Network was particularly interested in the government's support for coal power plants and published many commentaries explaining why building new coal power plants was not a good idea. During March and April 2015, when government committees were in the process of discussing Japan's future energy mix, Kiko Network's main concern was with some influential members of committees who emphasized the use of coal-fired thermal power plants, which were more cost efficient than natural gas-fired plants but emitted roughly twice the amount of CO_2. The government's Basic Energy Plan, adopted in April 2014, explained that coal-fired thermal power was a key source of electricity supply that excelled in terms of fuel supply stability and cost advantage. The draft of its long-term energy mix forecast assumed that coal would account for 26% of the nation's total electricity generation in 2030. This plan ran counter to the moves in many other industrialized economies, which were seeking to cut back on coal-fired power generation as part of the fight against climate change. Even Japan's financial aid for developing countries to buy Japanese coal-fired power generation technology exposed the nation to growing international criticism.

In 2015, Kiko Network reported that Japan's power companies had plans to build 45 new coal-fired plants. If all of these plants were put into operation, their total annual CO_2 emission output would be equivalent to about 10% of Japan's overall emissions in 1990. Kiko Network collaborated with other NGOs outside Japan and published a report on the development of coal power plants in the major developed countries (Kiko Network 2015). Japan was the only developed country planning to build so many new coal-fired power plants in the near future.

The public

With increasing evidence of the occurrence of extreme weather patterns and events in Japan, more people have begun to think that climate change had begun. At the same time, many people are convinced that reducing emissions will be costly. A questionnaire survey conducted by World Wide Views in June 2015 involved more than 10,000 participants in 76 countries. Japan's Science and Technology Agency (JST) was a partner organization in the project and conducted the questionnaire in Japan. Some of the results from its final report (JST 2015) are shown in Table 5.3.

Compared with the world average, ordinary Japanese people seemed to be less concerned about climate change, even though Japan had experienced a wide variety of extreme weather patterns in the last several years. Japanese people also perceived the efforts to fight climate change differently, in that a large majority of the people of Japan considered climate change mitigation action to be a threat to their

Table 5.3 Results of a questionnaire survey under the World Wide Views project

	Options (not all options are listed here)	World average	Japan
1. How concerned are you about the impacts of climate change?	a. Very concerned	79%	44%
	b. Concerned	19%	50%
2. In your opinion, measures to fight climate change are	a. Mostly a threat to our quality of life	27%	60%
	b. Mostly an opportunity to improve our quality of life	66%	17%
3. How should the world deal with exploration for new fossil fuel reserves?	a. Stop exploration for all fossil fuel reserves	45%	29%
	b. Only stop exploration for coal	17%	14%
	c. Continue all exploration	23%	39%

quality of life, whereas the majority global response was that mitigation actions represented an opportunity to improve quality of life. The reasons behind this unique perception of climate change were not evident from the survey, but they may include Japanese society's unwillingness to discuss climate change since the 11 March earthquake and the resulting four years of general neglect of the topic of climate change. This period represented a phase when the general public did not seem to have adequate information on how the rest of the world was making progress towards tackling the climate change issue.

Summary

The 11 March 2011 earthquake and the subsequent Fukushima Daiichi nuclear power accident surely had a great impact on Japan's climate change policy. People's awareness of nuclear energy policy dramatically increased after the event and did not fade for the next three or four years. This environment made it difficult to discuss climate change policy. Those who argued for the rehabilitation of nuclear power plants emphasized the threats posed by climate change to gain public support for the use of nuclear power. Therefore, discussions of climate change policy were inevitably intertwined with those of nuclear power, which remained controversial during this period

The only benefit Japan gained from the nuclear accident, in terms of climate change policy, was a strong push for the use of renewable energy. Renewables such as solar and wind power (excluding large-scale hydropower) supplied merely 1% of electricity before 2011. This rate was considerably lower than those in many other countries and even less than those in many other developing nations. Renewable energy finally gained governmental support, but only after the nuclear power plants were shut down. Under these extraordinary circumstances, people's awareness of the climate change issue receded. The first commitment period of the Kyoto Protocol ended in 2012, but most Japanese people seemingly were either uninterested or uninformed of the situation. There had been an increasing number of extreme weather events and patterns in Japan, including hotter summers and stronger typhoons, but the mass media did not explain that these extreme weather events could possibly be related to climate change.

Japan's GHG emission reduction target for 2030 was only determined by the Japanese government after its energy policy for 2030 was established. There was no discussion of how great an emission reduction was needed at the global level or of what Japan's reduction should be from an equity perspective. This decision-making process reveals a mind-set in which climate change policy was viewed as an energy issue rather than an environmental issue. In November 2014, at the time the United States and China jointly announced their emission reduction targets, many Japanese stakeholders were surprised to see these two countries taking the lead, because they had long considered them to be the two biggest laggards in terms of climate change policy. This indicates that people in Japan were generally unaware of important changes that had been occurring outside the country concerning climate change policies.

Japan's GHG emission reduction target assumes that nuclear power plants will be fully operational by 2030 and will provide about 22% of the total electricity supply. Some nuclear power plants slowly returned online in 2015, but it is still uncertain whether the Japanese public is ready to accept all of its nuclear power plants operating at the same level as that before the March 2011 accident. Two political parties, the DPJ and LDP, ruled between 2011 and 2015, but the two parties differed little in their approach to Japan's response to climate change. Former Prime Minister Hatoyama's 25% reduction target was not typical of the DPJ's overall position. Both parties regarded sustainable economic growth as Japan's top priority, rather than taking a lead in global issues, including climate change.

A change was observed, however, in the industrial sector. Positions on climate change became more diverse, depending on the various industries and business sectors. Energy-intensive industries such as power companies and steel companies remain the most reluctant to embrace GHG emission reduction targets. On the other hand, an increasing number of business organizations and companies appear to be more open to viewing the climate change problem as a new business opportunity.

Note

1   Sato, Katsuaki (2011) *Matsuo Basho* [Basho Matsuo]. Tokyo: Hitsuji Shobo.

References

Climate Action Tracker (CAT) (2015) *Japan's Proposed INDC "Inadequate" and Opposite to Its G7 Commitment*, 9 June 2015. Available online at: http://climateactiontracker. org/news/208/Japans-proposed-INDC-inadequate-and-opposite-to-its-G7-commitment. html (accessed 31 July 2015).

E3G (2015) *G7 Coal Scorecard – International Interest*. Available online at: http://www.e3g.org/news/media-room/g7-coal-scorecard-international-interest (accessed 30 January 2016).

Energy – Environment Council (2011) *Costo to kensho iinkai houkokusho* [Report from Costs Verification Committee] Energy – Environment Council, 19 December 2011.

Energy – Environment Council (2012a) *Enerugi Kankyo no Sentakushi ni kansuru Torongata Yoron Chosa Chosa Houkokusho* [Report on a deliberative polling survey concerning options on energy and environment] distribution material for the 12th Meeting of the Energy – Environment Council, 4 September 2012.

Energy – Environment Council (2012c) *Kakushinteki Energugi Kankyo Senryaku an* [Strategy for Innovative Energy and Environment (draft)] distribution material for the 14th Meeting of the Energy – Environment Council, 14 September 2012.

Government of Japan (2015) *Submission of Japan's Intended Nationally Determined Contibution (INDC)*. Available online at: http://www4.unfccc.int/submissions/INDC/Published%20Documents/Japan/1/20150717_Japan's%20INDC.pdf (accessed 30 January 2016).

The Guardian (2011) *Japan to scrap nuclear power in favour of renewables*. 10 May 2011. Available online at: https://www.theguardian.com/environment/2011/may/10/japan-nuclear-renewables (accessed 10 August 2016).

Intergovernmental Panel on Climate Change (IPCC) (2013) *Climate Change 2013: The Physical Science Basis*, Summary for Policymakers. Geneva: IPCC.

Japan-CLP (2015) Policy Paper on the Mid-Term GHG Reduction Target of Japan, May 2015. Japan Climate Leaders' Partnership.

Japan Meteorological Agency (2011) *Tohoku-chiho taiheiyo oki jishin* [The 2011 off the Pacific coast of Tohoku Earthquake]. Available online at: http://www.data.jma.go.jp/svd/eqev/data/2011_03_11_tohoku/ (accessed 20 November 2015).

Japan Science and Technology Agency (JST) (2015) *World Wide Views on Climate and Energy*. Available online at : http://www.jst.go.jp/csc/deliberation/WWV2015/document.html (accessed 20 November 2015).

Keidanren (2015a) Aratana Enerugi Mikkusu no Sakuteini Mukete [Concerning decision on new energy mix], Keidanren Policy & Action, 6 April 2015.

Keidanren (2015b) Aratana Kikohendowakugumi no Kochiku ni muketa Teigen: Kyoto kara Pari, soshite Jisedai no Chikyu he. [A Proposal for Structuring a New Climate Change Framework: From Kyoto to Paris, and to the Earth in the Next Generation], Keidanren Policy & Action, 11 September 2015.

Kiko Network (2015) *Press Release*, 25 November 2015. Available online at: http://www.japan-press.co.jp/modules/news/index.php?id=8759 (accessed 22 December 2015).

Kuramochi, Takeshi (2014) *GHG Mitigation in Japan: An Overview of the Current Policy Landscape*, Working Paper, World Resources Institute and Institute for Global Environmental Strategies.

Ministry of Economy, Trade and Industry (METI) (2014) *Basic Energy Plan*. Available online at: http://www.enecho.meti.go.jp/en/category/others/basic_plan/pdf/4th_strategic_energy_plan.pdf (accessed 20 November 2015).

Ministry of the Environment (MOE) (2013) *Statement by Nobuteru Ishihara, Japanese Minister of the Environment, at COP19/CMP9*. Available online at: http://www.env.go.jp/en/earth/cc/statement131120.pdf (accessed 30 January 2016).

Ministry of the Environment (MOE) (2014a) *Japan's National Greenhouse Gas Emissions in Fiscal Year 2013 (Preliminary Figures)*. Available online at: http://www.env.go.jp/en/headline/2132.html (accessed 22 December 2015).

Ministry of the Environment (MOE) (2014b) *Statement by Yoshio Mochizuki, Minister of the Environment of Japan, at COP 20*. Available online at: http://www.env.go.jp/en/earth/cc/cop20_statement_eng.pdf (accessed 30 January 2016).

Ministry of the Environment (MOE) (2014c) *Japan's Progress in Developing its Intended Nationally Determined Contribution*. Available online at: http://www.env.go.jp/en/earth/cc/jpdindc.html (accessed 30 January 2016).

Ministry of the Environment (2015a) New Mechanisms Information Platform website. Available online at: http://www.mmechanisms.org/e/ (accessed 30 January 2016).

Ministry of the Environment (MOE) (2015b) *Nippon no Onshitsu Koka Gasu Haishutu-ryo no Kekka* [Japan's Greenhouse Gas Inventory]. Available online at: http://www.env.go.jp/earth/ondanka/ghg/index.html (accessed 30 January 2016).

Nippon Keizai Shimbun (2015) "*Ondanka Gasu Sakugen no Shusho Shiji, Sanpou ni Hairyo*" [Prime Minister's Instruction on GHG Emission Reduction, Consideration of Three Instructions], 2 May 2015.

Oshima, Kenichi (2014) "On the Economy of Nuclear Power: Calculating the Actual Total Unit Cost of Power Production," in Hidenori Niizawa and Toru Morotomi eds., *Governing Low-Carbon Development and the Economy*. Tokyo: United Nations University Press, 235–249.

Prime Minister's Office of Japan (PMO) (2012) *2011nen Tohoku Chihou Taiheiyou Oki Jishin nitsuite* [A Report on East Japan Earthquake in 2011], 27 November 2012 (in Japanese).

Reuters (2011) *Japan Says Nuclear Policy Must Be Reviewed from Scratch*, 10 May 2011, Reuters. Available online at: http://www.reuters.com/article/us-japan-politics-pm-idUS-TRE7491SC20110511 (accessed 22 December 2015).

Stefanini, Sara (2015) "Merkel Convinces Canada and Japan on CO2: Germany Browbeats Canada and Japan into Joining Broad G7 Pledge to Cut Emissions," *Politico*, 8 June 2015. Available online at: http://www.politico.eu/article/germany-canada-japan-emissions-pledge/ (accessed 15 September 2015).

United Nations Environment Programme (UNEP) (2014) *The Emission Gap Report*, A UNEP Synthesis Report. Nariobi: UNEP.

United Nations Framework Convention on Climate Change (UNFCCC) (2011) Decision 1/CP.17, Establishment of an Ad Hoc Working Group on the Durban Platform for Enhanced Action, FCCC/CP/2011/9/Add.1.

United Nations Framework Convention on Climate Change (UNFCCC) (2013) Further Advancing the Durban Platform, Decision 1/CP.19, FCCC/P/2013/10/Add.1.

United Nations Framework Convention on Climate Change (UNFCCC) (2014) Decision 1/CP.20, Lima Call for Climate Action, FCCC/CP/2014/10/Add.1.

United Nations Framework Convention on Climate Change (UNFCCC) (2015a) Synthesis Report on the Aggregate Effect of the Intended Nationally Determined Contributions, Note by the Secretariat, FCCC/CP/2015/7.

United Nations Framework Convention on Climate Change (UNFCCC) (2015b) Paris Agreement, FCCC/CP/2015/L.9.We Mean Business (2015). Available online at: http://www.wemeanbusinesscoalition.org/ (accessed 30 January 2016).

World Nuclear Association (2015) *Nuclear Power in Japan*, Updated April 2015. Available online at: http://www.world-nuclear.org/info/Country-Profiles/Countries-G-N/Japan/ (accessed 30 January 2016).

Yamaji, Kenji (2015) *How Ambitious is the GHG Reduction Target of Japan?* ICE Forum meeting conference paper, 26 July 2015.

6 Conclusion

As we saw in the poems that introduced the previous chapters, some haiku indicate the time of year with a "season word," or *kigo* in Japanese. Representation of the seasons has traditionally been important in Japanese culture, including in tea ceremonies, flower arrangements, and poetry. Now that the seasons themselves have started to change as a result of global climate change, things that have not been traditionally observed are starting to become a reality in Japan. Climate change will affect Japan not only physically and economically, but also culturally and intellectually. The next generation in Japan may begin to sense Japan's four seasons in a way completely different from how they were sensed by their ancestors.

Meanwhile, responses of the Japanese people as well as the Japanese government to the climate change problem over the last three decades did not seem to care about the likely impact of climate change to Japan. This chapter is a concluding chapter, in which key findings in previous chapters are summarized, and prospects for the future are envisaged.

Key findings

Observations of Japan's decision-making in addressing the climate change problem highlighted several features that are typical of Japan's decision-making in general but unique to Japan. In addition, some other characteristics were observed that are specific to the climate change issue.

Japan framed climate change as an economic and energy problem

The climate change problem was framed as an economic as well as an energy problem throughout Japan's three decades of climate change policy-making. Each time Japanese stakeholders discussed Japan's emission targets, the discussions were framed from the perspective of "what can be done" to reduce greenhouse gas (GHG) emissions, not "what should be done" to mitigate climate change. People were not interested in the extent to which emissions needed to be reduced to achieve a safe level, or even in how a "safe level" should be defined in the first place. Debates on the global long-term goal of 2 °C, or on the notion of carbon budgets, seldom occurred.

This does not mean that Japanese decision-makers denied the science of climate change. Rather, they seem to have considered it meaningless to discuss goals that were either too idealistic or that they perceived as unachievable. Climate change deniers did exist in the general public, but they remained a small minority. Most Japanese people generally trusted the science underlying climate change and even perceived that climate change seemed to be occurring, but they still had little interest in taking any action by themselves. In the early years of the time period covered in this book, many Japanese thought that the adverse impacts of climate change would occur only in countries far from Japan or in developing countries – particularly small island states and low-lying countries. In the later years, as the adverse impact of climate change gradually began to become visible in Japan, Japanese people generally accepted these extreme weather events in the same way they would other natural hazards, such as earthquakes and traditional typhoons, which could not be avoided through human effort.

This mind-set toward climate change can be explained by several fundamental observations during the three decades of decision-making. First, Japanese people are generally receptive to natural changes. As illustrated in the haiku, people in Japan enjoy seasonal changes, but they do not consider human-induced seasonal changes to be plausible or controllable. Therefore, discussions related to the long-term targets of climate change policy, such as debates on the 2 °C temperature limit, were not generally acknowledged by ordinary Japanese citizens.

On the other hand, Japanese people were well versed with respect to topics related to energy conservation and energy efficiency. Given their experiences with the local pollution and oil crises of the 1950s, 1960s, and 1970s, people were well aware of the importance of conserving energy. People's experiences had taught them that environmental and natural resource issues could be overcome through technological innovation and diffusion. These experiences trained Japanese people to see environmental problems from either an energy- or technology-related perspective.

In terms of economics, almost all debates on Japan's climate policy have been framed as debates on economic policy. Japanese stakeholders have been concerned almost exclusively about the cost side of the mitigation debate, not about how many of the adverse effects of climate change could be avoided through the timely implementation of emission reduction policies.

US–Japan bilateral relations were viewed as the core of Japan's foreign policy, which influenced Japan's multilateral diplomacy on climate change

Japan's framing of the climate change problem as an economic and energy problem also affected its climate change diplomacy. Whereas some countries, particularly those in the EU, tended to see climate change as a foreign policy issue and were eager to take the lead in negotiations under the United Nations Framework Convention on Climate Change (UNFCCC), Japan was not interested in taking a lead in multilateral forums. Even though most Japanese at least generally understood that climate change was a global threat, many considered Japan to be a small country and felt that it alone could not change the growing trend of GHG emissions at the global level. Unlike some other countries that had GHG emissions as small as, or smaller than, those of Japan, Japan did not reach out to other countries to convince them to alter the growing trend of GHG emissions.

Japan also did not perceive climate change as a diplomatic issue, because it was not interested in the equity dimension of the problem as it related to developed and developing countries. Whereas developing countries consistently asserted that most of the current global warming resulted from past GHG emissions from developed countries and that developed countries should be responsible for their past emissions, this issue was rarely raised by Japanese decision-makers. Conversely, some Japanese stakeholders (particularly those in the industrial sector) framed the equity issue with respect to fairness among industries in Japan, the United States, and the EU. They insisted that Japanese industries were most disadvantaged relative to industries in other countries, particularly during the first commitment period of the Kyoto Protocol (2008–2012), because Japan's 6% emission reduction target was perceived as an extra burden on their production activities.

The Japanese government, on the other hand, always respected the US position on climate change. During negotiations leading up to the Kyoto conference in 1997, Japan's proposal on quantified limitation and reduction objectives (QEL-ROs) allowed countries with relatively high population growth rates (e.g., the United States) to set less stringent emission reduction targets than other countries. In the following negotiations, Japan generally followed the Umbrella Group's position. Japanese bureaucrats argued that Japan needed to respect much of the US position because it was the world's largest GHG emitter and any climate change regime would not be effective without its participation. Although this argument was true, it was more likely that Japanese decision-makers viewed Japan's alliance with the United States as the core of Japanese foreign policy. Japanese politicians and bureaucrats in various ministries have long had many bilateral relations with US politicians and other government officials and generally wished to continue to maintain a good relationship with the United States. In that sense, climate change was only one of a variety of issues between the two countries.

The only diplomatic avenue that was effective for pressuring Japan to take climate-friendly position was the G7/8 summit process. The summit has provided the Japanese prime ministers and the government with an opportunity to promote a good image of internationalists. As Hugo Dobson correctly argued in his precious work on Japan and G7/8 summit (Dobson 2004: 188), the summit is symbolic of Japan's position in the world and in this regard it may well be of more importance to the Japanese government and its people than to any other participating nations. Moreover, as the summit meetings are attended by several key European nations including Germany, France, and the United Kingdom, agenda on climate change is often likely to be inclined towards the EU position. Being the host of the Toyako G8 summit in 2008 surely affected Japan's position on long-term GHG emission target.

Japan did not frame the climate change problem from a justice perspective

On the basis of the two previous conclusions (Japan saw climate change mainly as an economic problem, and it did not view it from a justice and equity perspective between developed and developing countries), we can see why Japanese policy-makers tended to confine their discussion on Japan's emission reduction targets within debates on marginal abatement cost. Comparison of marginal abatement costs across countries was one way to compare energy efficiency in an economy. Marginal abatement costs do not reflect the equity dimension of the problem between rich and poor countries – an issue of great concern for most developing countries. Japan's position to avoid any discussion of the justice or equity aspects of climate change may have limited its opportunities to play any important roles in the international negotiations.

In a similar manner, Japanese policymakers were not interested in discussing the climate change problem from the viewpoint of social justice at domestic level. Environmental and social policies could be introduced hand-in-hand as there are common features between the two (Fitzpatrick 2014), but climate change policies were segregated from social policy in Japan.

Japan's politicians were unable to consider climate mitigation as a long-term strategy

Examination of the three decades of history related to climate change policy-making in Japan has shown that Japanese politicians were incapable to tackling climate change by seeing climate mitigation policy as a long-term strategy for Japan. Climate change could have been perceived as a reason to become more independent of fossil fuel imports, set higher energy-efficiency standards that could work to the advantage of Japanese manufacturers, and convince other countries to follow suit, as did political leaders in many other countries. Japanese prime ministers' terms were generally much shorter than those of leaders in other industrialized countries. The situation was even worse for the environment ministers, because 33 politicians experienced the position of environment minister in 28 years between 1988 and 2015. This was clearly a defect (at least in this case) of the Japanese political system if people had hoped their political leaders would consider the country's long-term future. Prime ministers who did not last more than a year were too busy putting their efforts into winning elections upfront and barely had time to create personal relationships with heads of state in other countries, or to construct policies that might be costly in the short term but that would bear fruit in the long term. Meanwhile, the prime ministers who enjoyed relatively long terms, such as Noboru Takeshita and Junichiro Koizumi, were not particularly pro-environment at the outset of their respective administrations, but generally took more pro-environment positions on climate change policy as their terms progressed.

There were other political leaders who worked for greener policies from the beginning of their careers. These politicians occasionally made important decisions that affected the country's destiny. Good examples include Sukio Iwatare, who strongly recommended hosting COP3 in Japan; Ryutaro Hashimoto, who instructed the governmental officials to agree to a 6% reduction target in the first commitment period of the Kyoto Protocol; and Yoriko Kawaguchi, who was an able negotiator during COP7. Without the support and efforts of these political leaders, officials from the Ministry of Environment (MOE) would not have been able to convince other stakeholders to push for a climate change policy in Japan. Unfortunately, these pro-environment political leaders were in a small minority in the Japanese political landscape.

Japan's bureaucracy used to hold decisive power, but more diverse stakeholders gradually began to participate in decision-making over time

As noted in Chapter 1, Japan has been well known for the strong decision-making power of bureaucrats. Ministries held great power to influence the perceptions of political leaders and build ties with industries. Any evaluation of Japan's climate change policy-making in early years, such as in the 1980s and 1990s, would lead to a similar conclusion: ministries – particularly MOE, Ministry of Economy, Trade and Industry (METI), and Ministry of Foreign Affairs (MOFA) – were the dominant players in the government in climate change policy-making. However, in about the mid-2000s, when international negotiators were searching for a post-Kyoto regime, different types of stakeholders began to show an increased willingness to be part of the decision-making process.

First and foremost among these was Japanese industry. Many industries were unhappy with the outcomes of COP3 and with the 6% reduction target under the Kyoto Protocol. Therefore, Japanese industries considered it indispensable to influence the Japanese government from the early stages of negotiations of the post-Kyoto framework. Nevertheless, even among industry and business groups, positions gradually diversified. Businesses that saw climate change as a new investment opportunity were more supportive of climate change policies than were other, more energy-intensive, industries that saw climate change policy as an obstacle.

Local governments also became more involved. They sometimes were more ambitious than the national government in implementing pollution abatement policies and other environmental policies, because these policies were supported by the local community. Local governments such as the Tokyo Metropolitan Government played a central role in introducing a cap and trade scheme in Japan. Other stakeholders such as environmental nongovernmental organizations (NGOs), citizens' groups, and researchers also began to become more engaged in decision-making.

This power shift from government bureaucrats to other stakeholders in Japan was observed not only in the climate change arena but also in Japanese policy-making in general. The iron triangle – the strong informal bonds among Liberal Democratic Party (LDP) faction leaders, bureaucrats of major ministries, and big businesses – slowly dissolved in Japan throughout the time period covered in this book, particularly as scandals and corruption became more evident ("The Man Who Remade Japan" 2006).

Lack of public participation

Gazing over more than two decades of climate change policy making in Japan, very little, if any, evidence existed that showed signs of public engagement. Most ordinary citizens are aware of climate change, but they do not have willingness to take action. The government, as a rule, allows about one month for the public to send comments to policy drafts before finalizing them, but it is mostly industry lobbies and environmental NGOs that send comments, both of which are not effective enough to make substantial changes to the drafts. A large majority of ordinary citizens keep silent or are uninterested in the debate. This is very different from what has been observed in some countries in Europe, or in some localities in North America, where citizens' movements became foundations of the regions' climate change policy making. It is natural that Japanese politicians show little interest in climate change when the public itself is not concerned about the problem.

Future of Japan's climate change policy

Once the Paris Agreement was reached at COP21, countries were expected to implement policies and measures sufficient to reach their emission reduction targets for 2030. What is the likely future of Japan's response to the climate change problem?

First of all, Japan is likely to continue to frame the climate change issue as an economic and energy policy issue. Although the direction of Japan's future energy strategy was finally determined in July 2015, a great deal of uncertainty remains. Japan's electricity market is scheduled to be liberalized, beginning in April 2016. Local monopolies by gigantic power companies are supposed to end (or at least be reduced in size), and various companies are expected to supply cheaper electricity. From this respect, many small coal-fired power plants are now in the planning stages. A large portion of the Japanese business sector still considers renewable energy to be too expensive and aims to introduce more coal-fired power, which is considered to be both the cheapest source of energy and the most stable in terms of supply (Littlecott 2015).

MOE struggled to delay the construction of these new coal power plants by requesting the power companies to reconsider the plan to comply with the government's energy plan as well as Japan's INDC. Strong resistance of the companies continued to exist throughout the year 2015. Finally in February 2016, ministers of METI and MOE achieved a deal. The environment minister accepted constuction of new coal fire power plants, as long as the plants were to be installed with the best available technologies with the highest energy efficiency.

Nuclear plants are slowly coming back online, but it is uncertain how many of the existing nuclear plants being placed back into operation will be acceptable to the public. It will be even more difficult for the public to accept the construction of new nuclear plants. The government chose 22% of total electricity to be supplied by nuclear power as a target, but it has no alternative solutions in case the Japanese public will not accept the use of so much nuclear power.

Renewable energy is still considered to be too expensive and too unstable to be a reliable and large source of energy. Japan has a great deal of potential to produce geothermal energy, but not much of it has been utilized, primarily because of opposition from hotels and resort businesses in hot spring areas. The Japanese government sees an especially modest role for wind, which is projected to contribute only 1.7% of the total electricity supply by 2030. It remains uncertain whether the Japanese people's perception of energy will shift toward a more climate-friendly direction.

On 22 December 2015, the first Joint Council Meeting after COP21 was held to discuss the Global Warming Mitigation Plan, which is to be finalized by spring 2016. There was a sort of "vacuum" period after the first commitment period of the Kyoto Protocol ended in 2012. The Japanese government had not established any formal national plan to mitigate GHG emission after 2012, mainly because of the earthquake in 2011, but also because the government maintained that the Cancun emission target for 2020 was not a legally binding commitment. Although this legal interpretation was correct, it was also true that there was no political will on the part of Japanese politicians to implement any GHG emission reduction measures independent of the legal status of the Cancun target. Despite the heralded success of COP21 and the adoption of the Paris Agreement, many members of the Industrial Structure Council have emphasized that the emission reduction target for 2030 is not legally binding, just like the Cancun Agreement, so that Japan does not need to make an overwhelming effort to reach the target. Some other members from the Central Environment Council emphasized the significance of mentioning the 1.5 °C temperature target in Article 2, as well as a sentence in Article 4.1 of the Paris Agreement saying that countries should "achieve a balance between anthropogenic emissions by sources and removals by sinks of GHGs in the second half of this century," but the former group argued that the "balance" can be achieved by the final year of this century and not necessarily by 2050, so that Japan should do away with its 80% emission reduction target for 2050, a long-term target adopted by the cabinet in 2009 just after the L'Aquila G8 Summit. The former group also expressed concern about the position of the United States, worrying about a déjà-vu situation in which the United States would withdraw from the Paris Agreement if a Republican were to take over the administration in 2017, similar to what happened with the Kyoto Protocol after G.W. Bush became president in 2001. A common feature of these comments is that that the debates of the Joint Council continue to be dominated by members who consider emission reductions to have a negative effect on the Japanese economy. In that sense, the Paris Agreement appears to have had little effect on these members.

Conversely, some members of the Japanese business community attended COP21 and appear to have recognized that the winds of change have arrived. They appear to understand that carbon-intensive industries have become less attractive in terms of financial investment – for example, in the coal industry. Less carbon-intensive businesses will become more economically competitive, and reducing GHG emissions may not represent a cost for the economy.

Second, the US–Japan bilateral relationship will continue as the core of Japan's foreign policy because of the strong security and economic ties between the two countries. At the same time, some high-level Japanese bureaucrats and politicians and those in the industrial sector are not enthusiastic about following the Obama administration's lead on climate change policies. As noted above, this is partly because of Japan's experience at the time of the Clinton administration and the Kyoto Protocol, and also partly because of a belief among high-level Japanese decision-makers that Republicans value the bilateral relationship more than do the Democrats (Reizei 2014). That said, the result of the upcoming US presidential election in November 2016 will affect not only the position of the United States but also that of Japan on climate change. Japan might be obliged to start taking a more ambitious position on climate change if the Democrats continue to hold power beyond 2017 or if the new Republican administration is more willing to take a lead in climate change policies in this new era.

China will become more influential in the climate change regime under the UNFCCC in the near future. This is because of many factors, including the size of its GHG emissions, the increasing size of its economy, and its increasing awareness that it can and should make use of the climate change issue to improve its diplomatic image in the multilateral arena. Meanwhile, the China–Japan relationship, which has been strained in recent years, is not likely to improve in the near future. Japan passed a series of laws on security policy in 2015 to allow the Japanese government to send troops overseas in case of emergency, and this has not been well received by the Chinese government. Therefore, small conflicts are likely to continue to occur between the two countries, at least in the near future. That said, China's position on climate change will not generally affect Japan's position on climate change.

In addition, it appears unlikely that Japan will consider using the climate change issue as a means of elevating its diplomatic position at the multilateral level. From this perspective, a majority of Japanese stakeholders are likely to continue to emphasize Japan's relatively small portion of total global GHG emissions, a share that will continue to decline in the future. This will also mean that Japan will not be viewed as a major player in multilateral negotiations on climate change. It is unknown whether Japanese negotiators in the future will view this as a diplomatic risk for Japan. Japan will maintain its strong willingness to utilize the Joint Crediting Mechanism (JCM) as a way to secure emission credits from overseas, as well as build strong economic bilateral ties with developing countries through the sale of Japanese technologies. Japanese negotiators could become more aware of the importance of securing Japan's diplomatic position in multilateral negotiations if they were to discover that Japan's declining position within the UNFCCC arena negatively affected the JCM program.

It is unlikely that climate justice is a concept that will be understood by Japanese audiences, even in the future. Japanese people are still not particularly interested in seeing climate change as a part of the larger development issue. It may even be possible to go a step further and say that Japanese people are generally not well trained in participating in logical discussions of justice. For example, a TV program called "Justice with Michael Sandel" and another Sandel publication (Sandel 2009) surged in popularity in Japan in 2010. This fascination could reflect Japanese people's inexperience in the deep contemplation of justice or other ethical debates. This was however a one-time fashion and rapidly dissolved after some years. Thus, it is unlikely that their thinking on the concept of justice will change in regard to a global issue such as climate change.

Third, most of Japan's political leaders will continue to be unable to view climate change as a long-term strategy. Japan's political system is not likely to change dramatically in the near future, and Japan's prime ministers will continue to change more frequently than leaders in other countries. This means that future Japanese politicians will still have to consider domestic short-term (primarily economic) issues more closely than diplomatic or environmental long-term issues. Japanese politicians have also traditionally had trouble speaking English. This may have also contributed to their reluctance to take on climate change as a means to exert their political leadership, and it may also explain why they appear to be unaffected by climate change discussions outside Japan, which are primarily in English at large conferences. Japan has been putting more effort into English education, so there may be a gradual long-term change in this respect.

Finally, the increasingly diversified number of stakeholders in the area of climate change should have a positive effect on Japan's implementation of climate change policies. Conflicts among the key ministries (MOE, METI, and MOFA) will continue to a certain extent in the future, but the ministries will not have the same level of influence as they have had in past years. In general, there will be much less of a top-down management system, where the central bureaucracy holds the ultimate power to control politicians, industries, and local governments. The political structure will increasingly become more of a bottom-up system, with multilevel implementation of climate change policies. One reason for this prediction is that, as discussed earlier, some of the people involved in decision-making at the national level have already emphasized that the emission reduction target in the Paris Agreement is not legally binding, implying that Japan does not need to take the target seriously. This means that it will be less likely that the Japanese national government will develop a comprehensive climate change mitigation plan that is ambitious enough to reach the INDC, and that other entities will have to step in.

Another reason is that Japan's decision-making process is generally becoming more fragmented. Political leaders in the past 20 years – particularly those in the Democratic Party of Japan (DPJ) – have made great efforts to consolidate power so that they can be less reliant on bureaucrats. At the same time, the Japanese government's financial deficit has become increasingly serious, and politicians have also become less capable of enjoying vested interests in maintaining good relationship with industry. In addition, industries that were quite powerful at the time of rapid industrialization in the 1960s and 1970s are not as influential as they used to be. Emerging new businesses are more likely to gain benefits from the world of low carbon development. It will become more difficult for the industry and business groups to form one voice on climate change policies.

The national government's budgetary constraints have also led local governments to become increasingly independent. With limited financial resources, local governments need to consider ways to improve living conditions, create jobs, and make their regions attractive to increase population. Some local governments have succeeded in combining climate change policy with other local policies, such as cleaner and safer energy sources (i.e., renewable energy) and building communities that are resilient in the face of natural hazards (i.e., through adaptation). Many other local governments are now preparing to take similar actions.

Environmental NGOs in Japan are likely to maintain a similar level of influence in the upcoming years. The concepts of NGOs, volunteering, and donating are gradually being accepted by Japanese society, and those related to climate change have also grown in the last two decades. The trend toward voluntary action taken in the case of emergency due to extreme weather events is growing, but the number of NGOs that pressure the government to aim at more ambitious emission reductions has been stable in recent years.

Final words

The adverse impact of climate change will become more evident in the future. The change has already affected Japanese fisheries and agriculture to some extent, and people engaged in these sectors have already adopted adaptation measures to mitigate the damage. More people are likely to lose their houses and lives owing to increasingly severe typhoons, flooding, and landslides. In addition, extreme heat in the summer could further increase the number of people who will be hospitalized or killed by heatstroke.

It is ultimately up to the Japanese people how to react to the losses and damages resulting from climate change. They can, on one hand, start paying more attention to the problem and begin to take action. Changing lifestyles and consumption patterns is the first thing people could do immediately. Changing consumption patterns includes choosing low-carbon products; this would, in turn, cause industries and businesses to change their production patterns. More and more companies could benefit economically by selling low-carbon goods and services. As these business sectors grow, political leaders will be obliged to become more aware of, and engaged in, the climate change problem. They would then be more likely to support emission reduction actions at the domestic level, while simultaneously trying to engage in reducing emissions abroad.

Japanese people could, on the other hand, choose to ignore climate change. The negative effects of climate change could be reluctantly accepted by saying *shikatanai* (it can't be helped). It is my hope the Japanese people will notice things that they have lost in their traditional culture, such as the seasonal colors and the natural landscape, before it is too late to take action.

References

Dobson, Hugo (2004) *Japan and the G7/G8: 1975–2002*, Sheffield Centre for Japanese Studies/RoutledgeCurzon Series. London and New York: RoutledgeCurzon.

Fitzpatrick, Tony ed. (2014) *International Handbook on Social Policy and the Environment*. Cheltenham: Edward Elgar.

Littlecott, Chris (2015) Japan Coal Phaseout G7 Scorecard Country Profile, E3G Summary October 2015.

"The Man Who Remade Japan: But Now Junichiro Koizumi Is Stepping Down" (2006) *The Economist*, 14 September.

Reizei, Alihiko (2014) "Bei minshutou ga hannichi toiu golai" ["Misunderstanding that U.S. Republicans cherish Japan"], *Newsweek*, 21 January. http://www.newsweekjapan. jp/reizei/2014/01/post-619.php (accessed 20 November 2015).

Sandel, J. Michael (2009) *Justice: What's the Right Thing to Do?* New York: Farrar Straus & Giroux.

Annex 1

Chronicle of Japan's climate change policy

		Climate change-related events at international level	Climate-change related events in Japan	Prime minister of Japan (LDP, as long as otherwise mentioned)	Environment Minister	Non-climate-related events, domestic & international
1987	Jan.			Yasuhiro Nakasone	Toshiyuki Inamura	
	Feb.					
	Mar.					
	Apr.					
	May					
	Jun.					G7 Venice Summit
	Jul.					
	Aug.					
	Sep.	Villach workshop "Developing policies for responding to future climatic change"				
	Oct.					
	Nov.	Bellagio Policy workshop "Developing policies for responding to future climatic change"		Noboru Takeshita	Toshio Horiuchi	
	Dec.					
1988	Jan.					
	Feb.					
	Mar.					
	Apr.					
	May					

(Continued)

Chronicle of Japan's climate change policy (Continued)

	Climate change-related events at international level	Climate-change related events in Japan	Prime minister of Japan (LDP, as long as otherwise mentioned)	Environment Minister	Non-climate-related events, domestic & international
Jun.	Toronto Conference on the Changing Atmosphere				G7 Toronto Summit
Jul.					
Aug.					
Sep.					
Oct.					
Nov.	IPCC established				
Dec.					
1989 Jan.				Masahiro Aoki	Inauguration of G.H.W. Bush as President of the US
Feb.					
Mar.					
Apr.					
May					
Jun.			Sousuke Uno	Tatsuo Yamazaki	Tiananmen Square incident in China
Jul.					G7 Paris Summit
Aug.			Toshiki Kaifu	Mayumi Moriyama / Setsu Shiga	
Sep.					
Oct.					
Nov.	Noordwijk ministerial conference				

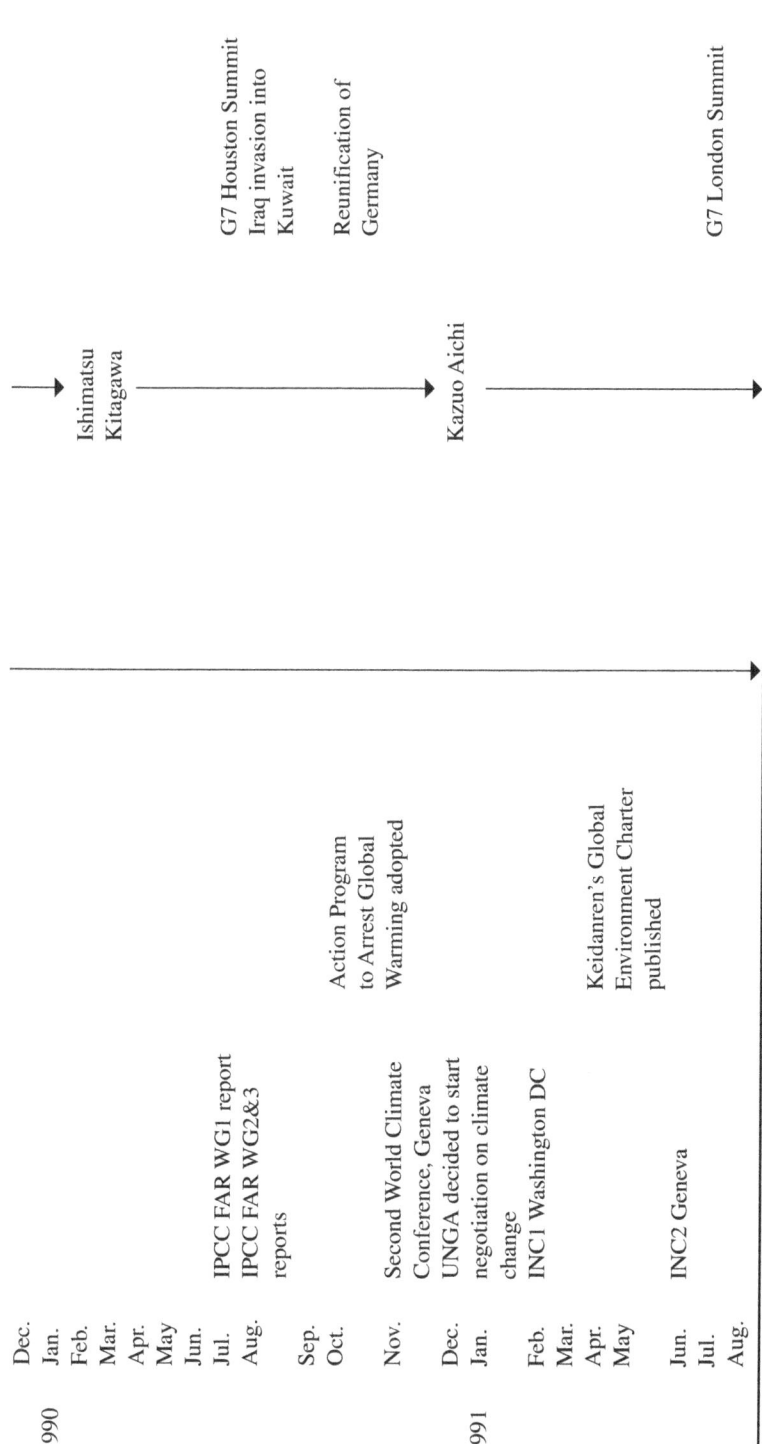

1990	Dec.			
	Jan.			Ishimatsu Kitagawa
	Feb.			
	Mar.			
	Apr.			
	May			
	Jun.			
	Jul.	IPCC FAR WG1 report		G7 Houston Summit
	Aug.	IPCC FAR WG2&3 reports		Iraq invasion into Kuwait
	Sep.			
	Oct.		Action Program to Arrest Global Warming adopted	Reunification of Germany
	Nov.	Second World Climate Conference, Geneva		
	Dec.	UNGA decided to start negotiation on climate change		
1991	Jan.	INC1 Washington DC		Kazuo Aichi
	Feb.			
	Mar.			
	Apr.		Keidanren's Global Environment Charter published	
	May			
	Jun.	INC2 Geneva		
	Jul.			G7 London Summit
	Aug.			

(Continued)

Chronicle of Japan's climate change policy (Continued)

	Climate change-related events at international level	Climate-change related events in Japan	Prime minister of Japan (LDP, as long as otherwise mentioned)	Environment Minister	Non-climate-related events, domestic & international
Sep.	INC3 Nairobi				Japanese government starts debate on Law on Cooperation in UN Peacekeeping Operations
Oct.					
Nov.			Kiichi Miyazawa	Shozaburo Nakamura	
Dec.	INC 4 Geneva				
1992 Jan.					
Feb.	INC5 New York	Council on Basic Environment Issues was established in LDP			
Mar.					
Apr.		Tokyo High-level Conference on Global Environment Issues			
May	INC5–2, New York, UNFCCC adopted				
Jun.	UNCED				
Jul.					G7 Munich Summit
Aug.					
Sep.					
Oct.					
Nov.					
Dec.	INC6 Geneva			Taikan Hayashi	
1993 Jan.					
Feb.					Inauguration of B. Clinton as President of the US

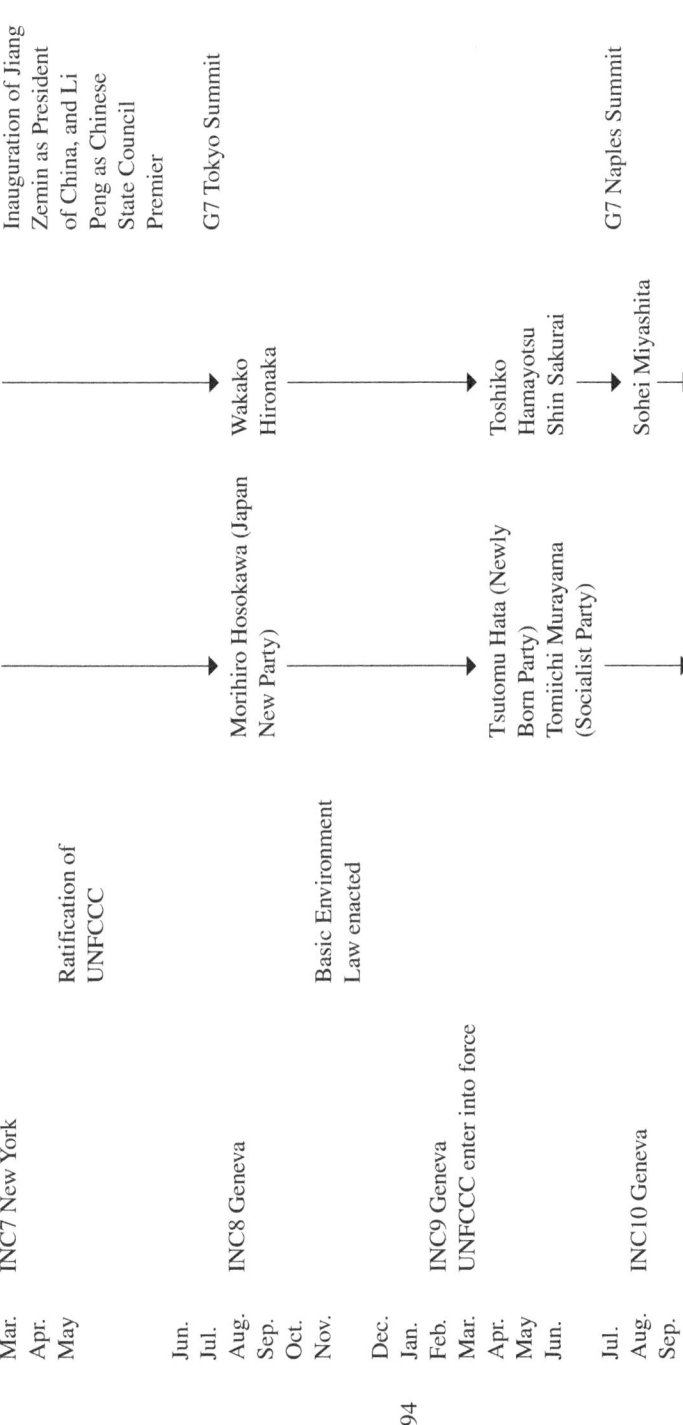

	Mar.	INC7 New York			Inauguration of Jiang Zemin as President of China, and Li Peng as Chinese State Council Premier
	Apr.				
	May		Ratification of UNFCCC		
	Jun.				
	Jul.				G7 Tokyo Summit
	Aug.	INC8 Geneva		Morihiro Hosokawa (Japan New Party)	Wakako Hironaka
	Sep.				
	Oct.				
	Nov.		Basic Environment Law enacted		
	Dec.				
1994	Jan.				
	Feb.	INC9 Geneva			
	Mar.	UNFCCC enter into force			
	Apr.			Tsutomu Hata (Newly Born Party)	Toshiko Hamayotsu
	May				Shin Sakurai
	Jun.			Tomiichi Murayama (Socialist Party)	
	Jul.	INC10 Geneva			Sohei Miyashita
	Aug.				
	Sep.				G7 Naples Summit

(Continued)

Chronicle of Japan's climate change policy (Continued)

	Climate change-related events at international level	Climate-change related events in Japan	Prime minister of Japan (LDP, as long as otherwise mentioned)	Environment Minister	Non-climate-related events, domestic & international
Oct.					
Nov.					
Dec.		The First Basic Environment Plan adopted			
1995 Jan.					Hanshin–Awaji Earthquake
Feb.					
Mar.	COP1, Berlin				Aum religious group's incidents
Apr.	COP1, Berlin, Berlin Mandate	Environment minister made a statement Japan's intention to host one of future COPs			
May					
Jun.					G7 Halifax Summit
Jul.					Murayama Danwa
Aug.				Tadamori Ohshima	
Sep.					
Oct.					
Nov.					

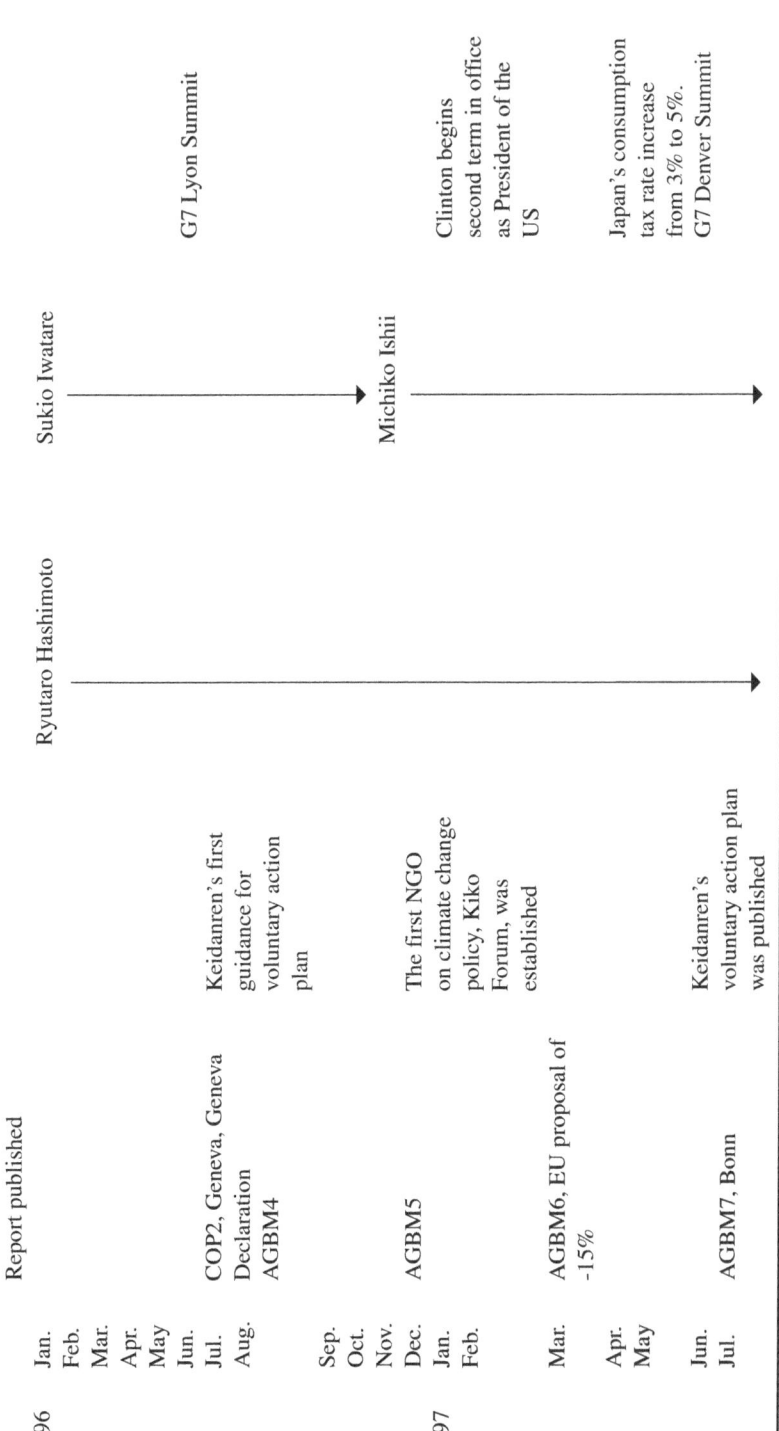

1996	Dec.	IPCC 2nd Assessment Report published		
	Jan.			
	Feb.			
	Mar.			
	Apr.			
	May			
	Jun.			
	Jul.	COP2, Geneva, Geneva Declaration	Keidanren's first guidance for voluntary action plan	G7 Lyon Summit
	Aug.	AGBM4		
	Sep.			
	Oct.			
	Nov.			
	Dec.	AGBM5	The first NGO on climate change policy, Kiko Forum, was established	
1997	Jan.			Clinton begins second term in office as President of the US
	Feb.			
	Mar.	AGBM6, EU proposal of -15%		
	Apr.			Japan's consumption tax rate increase from 3% to 5%.
	May			
	Jun.		Keidanren's voluntary action plan was published	G7 Denver Summit
	Jul.	AGBM7, Bonn		

Ryutaro Hashimoto

Sukio Iwatare

Michiko Ishii

(*Continued*)

Chronicle of Japan's climate change policy (Continued)

	Climate change-related events at international level	Climate-change related events in Japan	Prime minister of Japan (LDP, as long as otherwise mentioned)	Environment Minister	Non-climate-related events, domestic & international
Aug.				Hiroshi Ohki	
Sep.				↓	
Oct.	AGBM8, Bonn	Japan proposed emission reduction targets	↓	↓	
Nov.			↓	↓	
Dec.	COP3, Kyoto, adoption of Kyoto Protocol	Establishment of Global Warming Prevention Headquarter	↓	↓	
1998 Jan.			↓	↓	
Feb.			↓	↓	
Mar.			↓	↓	
Apr.			↓	↓	
May			↓	↓	G8 Birmingham Summit
Jun.		Guidelines of Measures to Prevent Global Warming adopted	↓	↓	
Jul.			Keizo Obuchi	Kenji Manabe	
Aug.			↓	↓	
Sep.			↓	↓	
Oct.			↓	↓	
Nov.	COP4, Buenos Aires, Buenos Aires Plan of Action	Law Concerning the Promotion of Measures to Cope with Global Warming enacted	↓	↓	

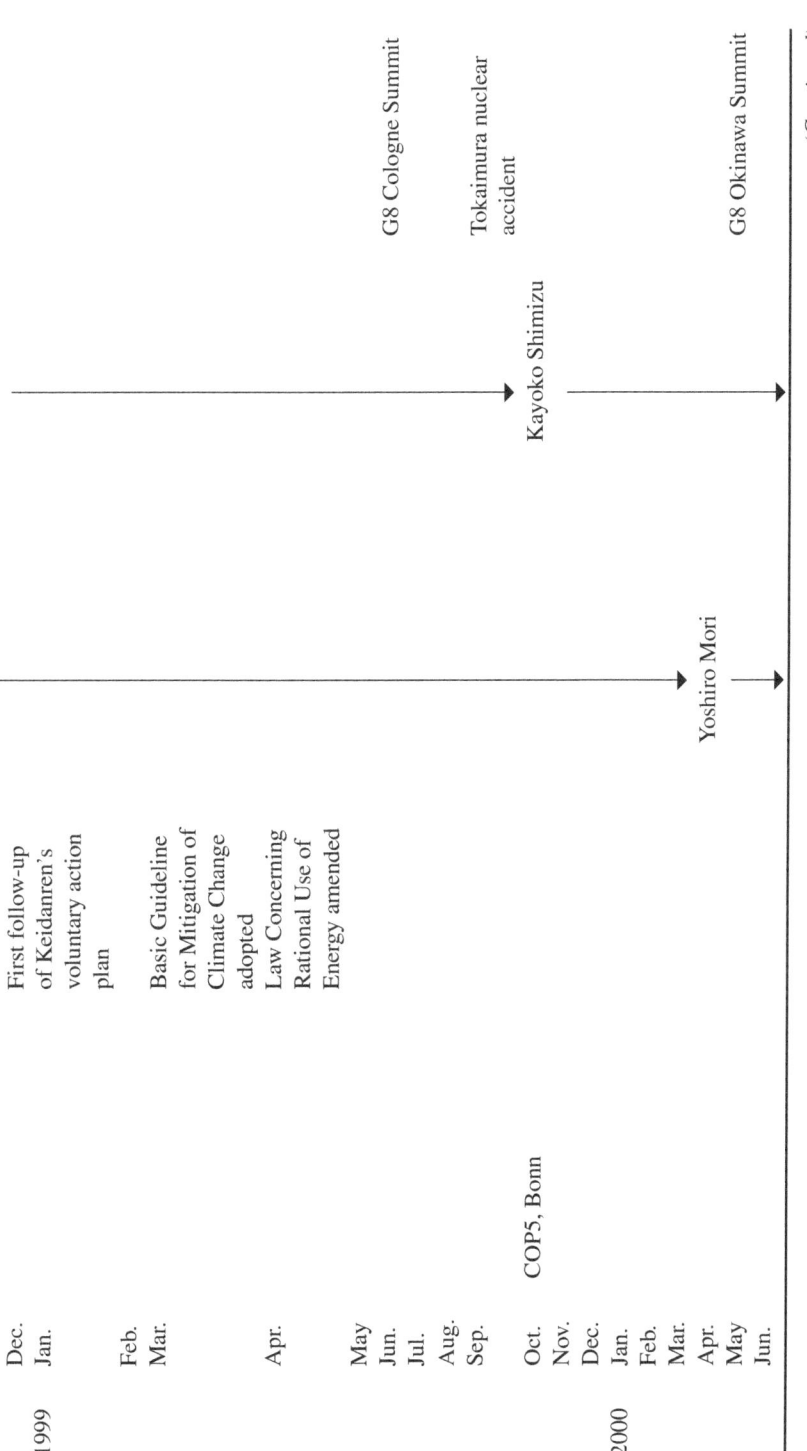

1999	Dec.		
	Jan.	First follow-up of Keidanren's voluntary action plan	
	Feb.		
	Mar.	Basic Guideline for Mitigation of Climate Change adopted	
	Apr.	Law Concerning Rational Use of Energy amended	
	May		
	Jun.		G8 Cologne Summit
	Jul.		
	Aug.		
	Sep.		Tokaimura nuclear accident
	Oct.	COP5, Bonn	
	Nov.		
	Dec.		
2000	Jan.		
	Feb.		
	Mar.		
	Apr.		
	May		
	Jun.		G8 Okinawa Summit

Yoshiro Mori

Kayoko Shimizu

(Continued)

Chronicle of Japan's climate change policy (Continued)

	Climate change-related events at international level	Climate-change related events in Japan	Prime minister of Japan (LDP, as long as otherwise mentioned)	Environment Minister	Non-climate-related events, domestic & international
Jul.				Yoriko Kawaguchi	
Aug.					
Sep.					
Oct.					
Nov.	COP6, Hague				
Dec.		The 2nd Basic Environment Plan adopted			
2001 Jan.	IPCC TAR WG1 report published	Environment Agency renamed to Ministry of the Environment			Inauguration of G. W. Bush as President of the US
Feb.	IPCC TAR WG2 report published				
Mar.	IPCC TAR WG3 report published; US announced its withdrawal from the Kyoto Protocol.				United States withdrew from the Kyoto Protocol.
Apr.					
May			Junichiro Koizumi		
Jun.					
Jul.	COP6–2 resume, Bonn				G8 Genoa Summit
Aug.					
Sep.					9.11 terrorists attack in the United States

Yoriko Kawaguchi

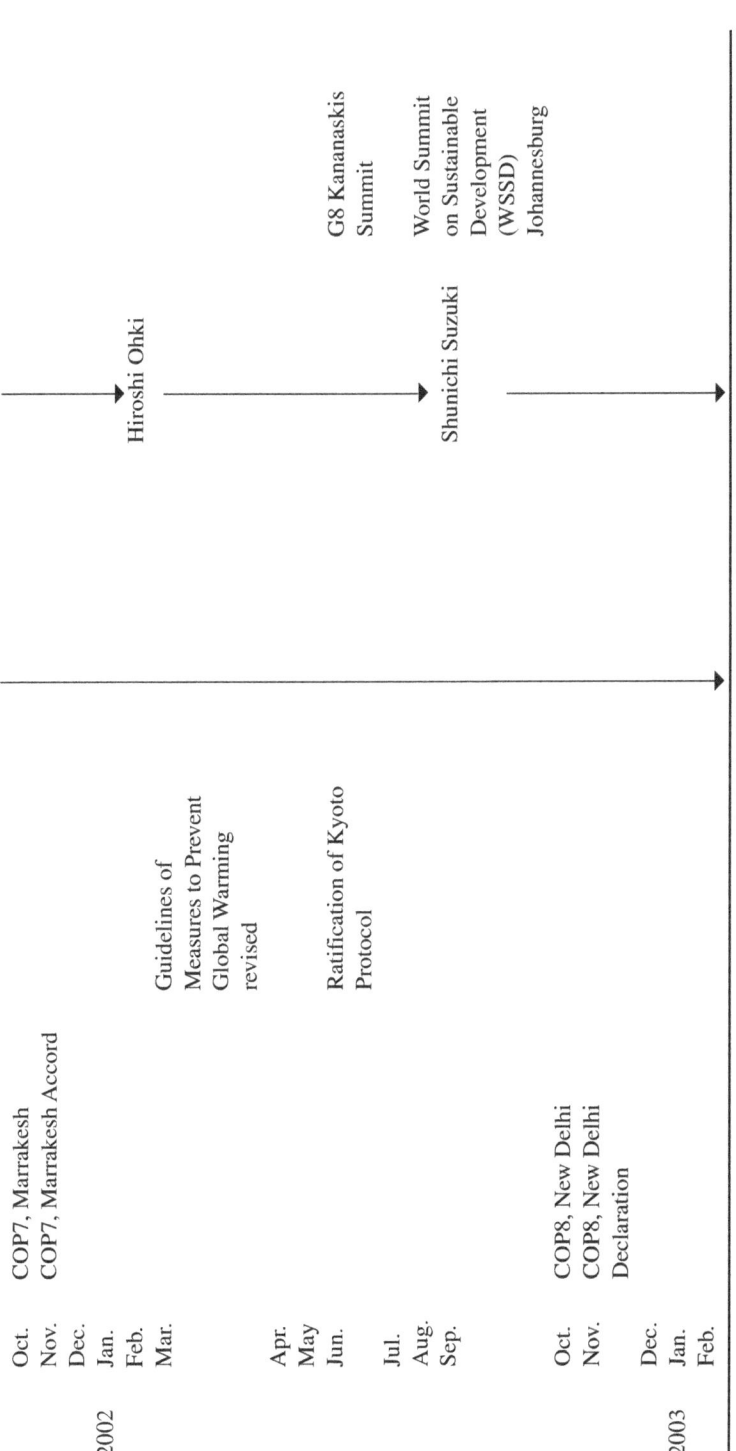

2002		
Oct.	COP7, Marrakesh	
Nov.	COP7, Marrakesh Accord	
Dec.		
Jan.		
Feb.		
Mar.	Guidelines of Measures to Prevent Global Warming revised	
Apr.		
May		
Jun.	Ratification of Kyoto Protocol	G8 Kananaskis Summit
Jul.		
Aug.		World Summit on Sustainable Development (WSSD) Johannesburg
Sep.		
Oct.	COP8, New Delhi	
Nov.	COP8, New Delhi Declaration	
Dec.		
2003		
Jan.		
Feb.		

Hiroshi Ohki

Shunichi Suzuki

(Continued)

Chronicle of Japan's climate change policy (Continued)

	Climate change-related events at international level	Climate-change related events in Japan	Prime minister of Japan (LDP, as long as otherwise mentioned)	Environment Minister	Non-climate-related events, domestic & international
Mar.					Inauguration of Hu Jintao as President of China, and Wen Jiabao as Chinese State Council Premier
Apr.					
May					G8 Evian Summit
Jun.					
Jul.		Industrial Structure Council published an interim report			
Aug.					
Sep.				Yuriko Koike	
Oct.					The first Basic Energy Plan adopted by the cabinet
Nov.					
Dec.	COP9, Milan				
2004 Jan.		Central Environment Council published an interim report			
Feb.					
Mar.					
Apr.					
May					
Jun.					G8 Sea Island Summit

Jul.		
Aug.		
Sep.		
Oct.		
Nov.	Russia ratified the Kyoto Protocol	
Dec.	COP10, Buenos Aires	
2005 Jan.	EU ETS started	G.W. Bush began second term in office as the US President
Feb.	Kyoto Protocol entered into force	
Mar.		
Apr.	Kyoto Protocol Target Achievement Plan	
May	JVETS (voluntary emissions trading scheme) started	
Jun.	Cool Biz Campaign started	
Jul.		G8 Gleneagles Summit Koizumi Danwa
Aug.		
Sep.		
Oct.		
Nov.	COP11, CMP1, Montreal	

(Continued)

Chronicle of Japan's climate change policy (Continued)

		Climate change-related events at international level	Climate-change related events in Japan	Prime minister of Japan (LDP, as long as otherwise mentioned)	Environment Minister	Non-climate-related events, domestic & international
2006	Dec.	COP11, CMP1,Montreal				
	Jan.	Establishment of AWG-KP				
	Feb.					
	Mar.					
	Apr.					
	May		The 3rd Basic Environment Plan adopted			
	Jun.					
	Jul.					G8 St Petersburg Summit
	Aug.					
	Sep.			Shinzo Abe	Masatoshi Wakaba-yashi	
	Oct.					
	Nov.	COP12, CMP2, Nairobi				
	Dec.					
2007	Jan.					
	Feb.	IPCC AR4 WG1 report				
	Mar.					
	Apr.	IPCC AR4 WG2 report				
	May	IPCC AR4 WG3 report	Abe's statement on "Beautiful Planet 50"			
	Jun.					G8 Heiligendamm summit
	Jul.					
	Aug.				Ichiro Kamoshita	

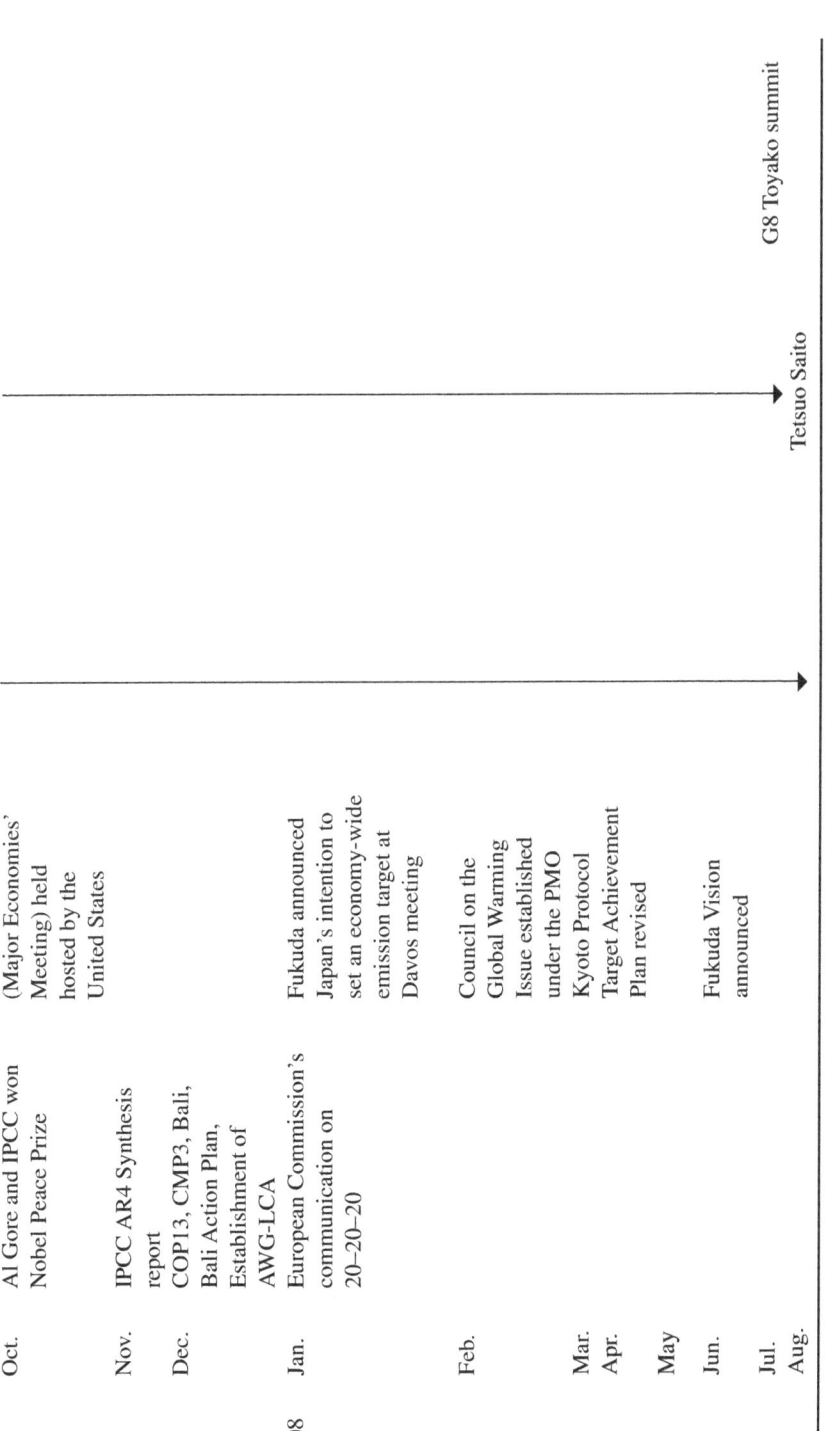

			Yasuo Fukuda	Tetsuo Saito
Sep.				
Oct.	Al Gore and IPCC won Nobel Peace Prize	The first MEM (Major Economies' Meeting) held hosted by the United States		
Nov.	IPCC AR4 Synthesis report			
Dec.	COP13, CMP3, Bali, Bali Action Plan, Establishment of AWG-LCA			
2008 Jan.	European Commission's communication on 20–20–20	Fukuda announced Japan's intention to set an economy-wide emission target at Davos meeting		
Feb.		Council on the Global Warming Issue established under the PMO		
Mar.				
Apr.		Kyoto Protocol Target Achievement Plan revised		
May				
Jun.		Fukuda Vision announced		
Jul.				G8 Toyako summit
Aug.				

(Continued)

Chronicle of Japan's climate change policy (Continued)

	Climate change-related events at international level	Climate-change related events in Japan	Prime minister of Japan (LDP, as long as otherwise mentioned)	Environment Minister	Non-climate-related events, domestic & international
Sep.			Taro Aso		Lehman Brothers bankruptcy
Oct.		Pilot-phase ETS started			
Nov.		Committee on the			
Dec.	COP14, CMP4, Poznan	Mid-term Emission Target established			
2009 Jan.					Inauguration of B. Obama as President of the US
Feb.					
Mar.					
Apr.					
May					
Jun.		Japan's emission reduction target for 2020 determined			The US House of Representatives passed Clean Energy and Security Act bill
Jul.					G8 L'Aquila summit
Aug.					

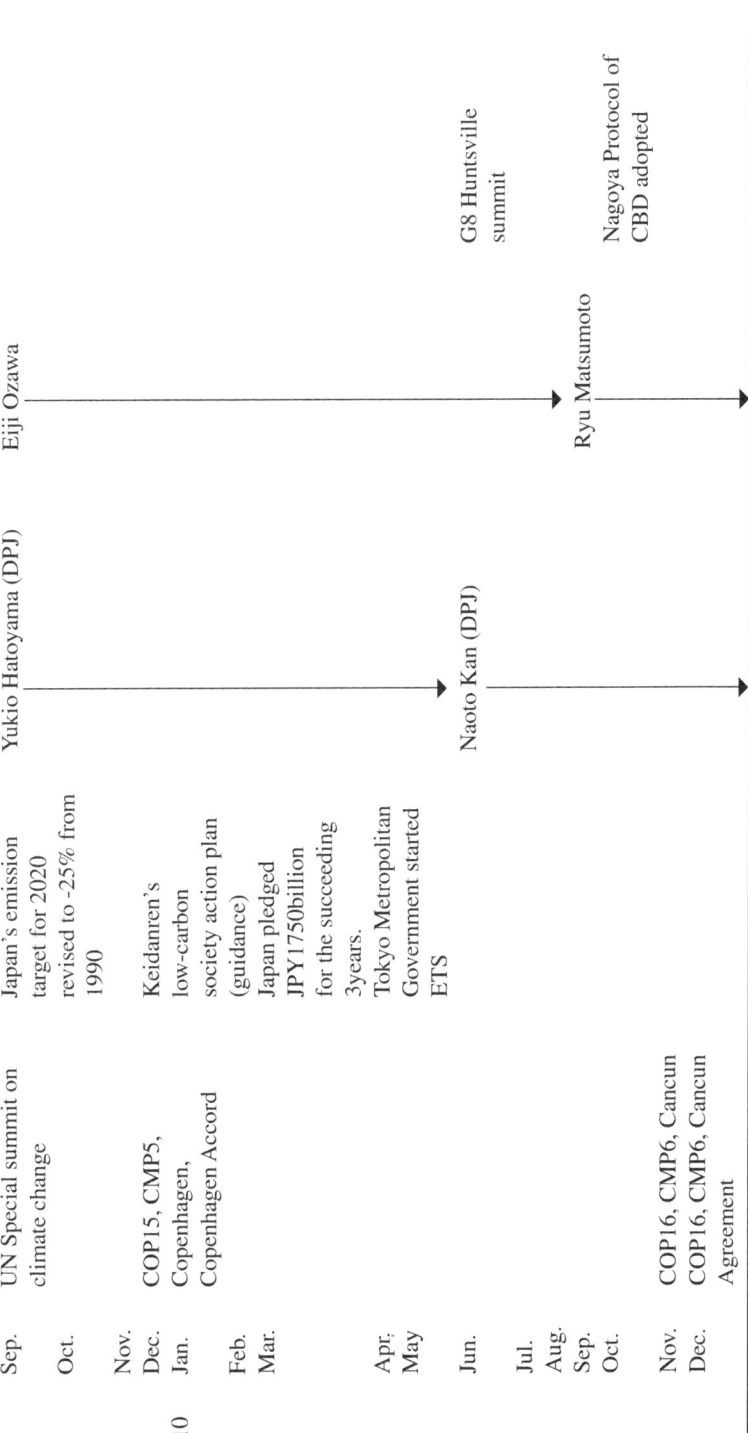

	Sep.	UN Special summit on climate change	Yukio Hatoyama (DPJ)	Eiji Ozawa
	Oct.	Japan's emission target for 2020 revised to −25% from 1990		
	Nov.			
	Dec.	COP15, CMP5, Copenhagen, Copenhagen Accord		
2010	Jan.	Keidanren's low-carbon society action plan (guidance)		
	Feb.			
	Mar.	Japan pledged JPY1750billion for the succeeding 3years.		
	Apr; May	Tokyo Metropolitan Government started ETS		
	Jun.		Naoto Kan (DPJ)	
	Jul.			
	Aug.			
	Sep.			G8 Huntsville summit
	Oct.			Ryu Matsumoto
	Nov.	COP16, CMP6, Cancun		Nagoya Protocol of CBD adopted
	Dec.	COP16, CMP6, Cancun Agreement		
2011	Jan.			

(Continued)

Chronicle of Japan's climate change policy (Continued)

	Climate change-related events at international level	Climate-change related events in Japan	Prime minister of Japan (LDP, as long as otherwise mentioned)	Environment Minister	Non-climate-related events, domestic & international
Feb.					
Mar.					March 11 Earthquake in Tohoku region
Apr.					
May					G8 Deauville Summit
Jun.		Energy-Environment Council was established			
Jul.		Special Law to Promote Renewable Energy enacted		Satsuki Eda	
Aug.					
Sep.			Yoshihiko Noda (DPJ)	Goshi Hosoda	
Oct.		Committee for Verification of Costs was set up			
Nov.	COP17, CMP7, Durban				
Dec.	COP17, CMP7, Durban Platform	Feed-in tariff scheme was introduced			
2012 Jan.					
Feb.					
Mar.					

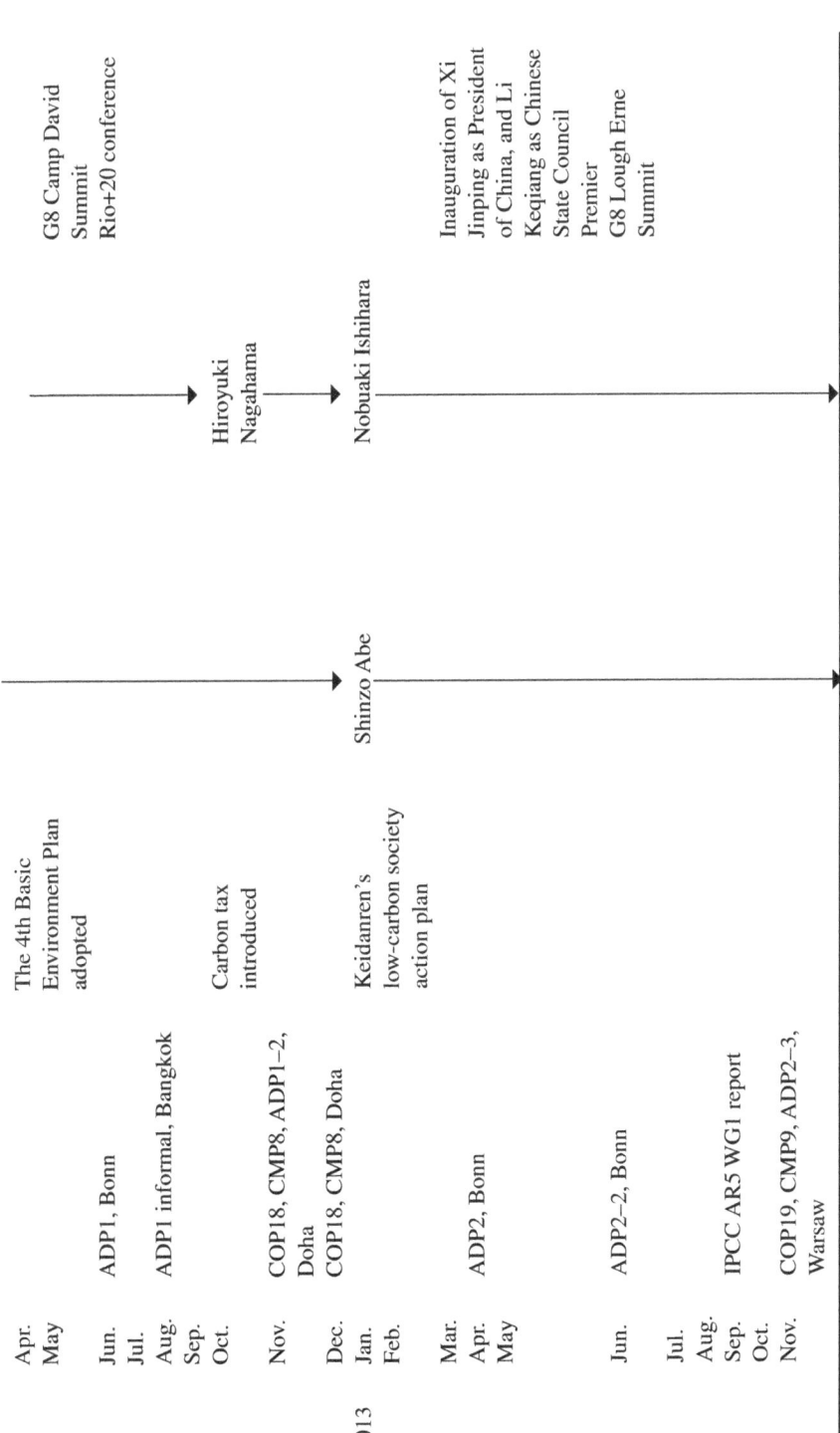

	Japan	Prime Minister / Environment Minister	International
Apr.			
May	The 4th Basic Environment Plan adopted		G8 Camp David Summit
Jun.	ADP1, Bonn		Rio+20 conference
Jul.			
Aug.	ADP1 informal, Bangkok		
Sep.			
Oct.	Carbon tax introduced	Hiroyuki Nagahama	
Nov.	COP18, CMP8, ADP1–2, Doha		
Dec.	COP18, CMP8, Doha		
2013 Jan.	Keidanren's low-carbon society action plan	Shinzo Abe → / Nobuaki Ishihara →	
Feb.			
Mar.			Inauguration of Xi Jinping as President of China, and Li Keqiang as Chinese State Council Premier
Apr.	ADP2, Bonn		
May			
Jun.	ADP2–2, Bonn		G8 Lough Erne Summit
Jul.			
Aug.			
Sep.	IPCC AR5 WG1 report		
Oct.			
Nov.	COP19, CMP9, ADP2–3, Warsaw		

(Continued)

Chronicle of Japan's climate change policy (Continued)

	Climate change-related events at international level	Climate-change related events in Japan	Prime minister of Japan (LDP, as long as otherwise mentioned)	Environment Minister	Non-climate-related events, domestic & international
2014					
Dec.					
Jan.					
Feb.					
Mar.	ADP2–4, IPCC AR5 WG2 report				
Apr.	IPCC AR5 WG3 report	Cabinet adopted a New Basic Energy Plan			Japan's consumption tax rate increase from 5% to 8%
May					G7 Brussels summit
Jun.	ADP2–5, Bonn				
Jul.					
Aug.					
Sep.				Yoshio Mochizuki	
Oct.	ADP2–6, Bonn				
Nov.	IPCC AR5 Synthesis report				
Dec.	US and China's INDC announced COP20, CMP10, ADP2–7, Lima				
2015					
Jan.					
Feb.	ADP2–8, Geneva				
Mar.					
Apr.					
May					

Jun.	ADP2–9, Bonn		G7 Elmau summit
Jul.			
Aug.	ADP2–10, Bonn	Adoption of long-term energy plan, Submission of Japan's INDCS	
Sep.			UN Sustainable Development Summit
		Tamayo Marukawa	
Oct.	ADP2–11, Bonn		
Nov.	COP21, Paris	Cabinet approval of Adaptation Plan	Terrorists attack in Paris
Dec.	COP21, CMP11, Paris, adoption of Paris Agreement		

Index